Springer Theses

Recognizing Outstanding Ph.D. Research

Aims and Scope

The series "Springer Theses" brings together a selection of the very best Ph.D. theses from around the world and across the physical sciences. Nominated and endorsed by two recognized specialists, each published volume has been selected for its scientific excellence and the high impact of its contents for the pertinent field of research. For greater accessibility to non-specialists, the published versions include an extended introduction, as well as a foreword by the student's supervisor explaining the special relevance of the work for the field. As a whole, the series will provide a valuable resource both for newcomers to the research fields described, and for other scientists seeking detailed background information on special questions. Finally, it provides an accredited documentation of the valuable contributions made by today's younger generation of scientists.

Theses are accepted into the series by invited nomination only and must fulfill all of the following criteria

- They must be written in good English.
- The topic should fall within the confines of Chemistry, Physics, Earth Sciences, Engineering and related interdisciplinary fields such as Materials, Nanoscience, Chemical Engineering, Complex Systems and Biophysics.
- The work reported in the thesis must represent a significant scientific advance.
- If the thesis includes previously published material, permission to reproduce this must be gained from the respective copyright holder.
- They must have been examined and passed during the 12 months prior to nomination.
- Each thesis should include a foreword by the supervisor outlining the significance of its content.
- The theses should have a clearly defined structure including an introduction accessible to scientists not expert in that particular field.

More information about this series at http://www.springer.com/series/8790

Lokesh Dhakar

Triboelectric Devices for Power Generation and Self-Powered Sensing Applications

Doctoral Thesis accepted by
National University of Singapore, Singapore

Springer

Author
Dr. Lokesh Dhakar
Department of Electrical and Computer Engineering
National University of Singapore
Singapore

Supervisor
Prof. Francis E.H. Tay
Department of Mechanical Engineering
National University of Singapore
Singapore

Additional material to this book can be downloaded from http://extras.springer.com.

ISSN 2190-5053 ISSN 2190-5061 (electronic)
Springer Theses
ISBN 978-981-10-3814-3 ISBN 978-981-10-3815-0 (eBook)
DOI 10.1007/978-981-10-3815-0

Library of Congress Control Number: 2017931527

Printed on acid-free paper

This Springer imprint is published by Springer Nature
The registered company is Springer Nature Singapore Pte Ltd.
The registered company address is: 152 Beach Road, #21-01/04 Gateway East, Singapore 189721, Singapore

Supervisor's Foreword

The progress of humanity has been fueled by the advent of technologies such as steam engine, electricity, transistors and internet. As we move forward, there is a continuous strive towards making our surroundings smarter and intelligent, which means we are looking at a future with myriad of sensors around us. One of the biggest challenges in this aspect to devise a sustainable way of powering these sensors. A potential strategy of achieving this is to design the sensor system in such a way that they are able to capture energy from their surroundings and convert it into usable form, which can be stored into batteries or capacitors. This process of capturing energy from surroundings is known as "Energy Harvesting". The energy available in the environment can be in various forms: solar, heat, acoustic or mechanical. This work specifically focuses on developing energy harvesting devices that convert mechanical energy available in the surroundings to electrical energy using triboelectric mechanism.

Triboelectric mechanism has been known for centuries in the form of static electricity. Recently, researchers have demonstrated a new application for this phenomenon—to convert mechanical energy into electrical energy. Lokesh has been involved in research of triboelectric mechanism since the scientific community began working in this area in 2012. His commitment towards research combined with hard work has resulted into this outstanding thesis. This work has been published in prestigious journals and conferences and has also been covered by many popular science media outlets. A part of this work published as a paper in the Journal of Micromechanics and Microengineering was selected as one of the top highlighted paper in the journal in 2014. He was also the outstanding oral paper award finalist for IEEE International Conference on Micro Electro Mechanical Systems 2015.

One of the most important factors which affect the performance of triboelectric mechanism based devices is the topography of the triboelectric layers. This work systematically studies the effect of micropatterned arrays on the performance of triboelectric devices. This provides important guidelines for researchers to design and pattern devices with the aim of improving their device performance.

Triboelectric mechanism based devices have enormous application in the area of wearable devices due to a wide choice of fabrication materials. The human skin has been proposed and demonstrated as one of the possible materials for biomechanical energy harvesting and sensing applications using triboelectric mechanism. Furthermore, a novel design for triboelectric mechanism based devices that is sensitive to both out-of-plane and in-plane mechanical stimulation has been proposed. Commercial viability of any technology depends on the scalable fabrication and economic production of devices. This work also explores and demonstrates roll-to-roll fabrication process as a method to fabricate triboelectric mechanism based devices in a scalable fashion.

This thesis work will be of interest to graduate students and researchers working in the area of mechanical energy harvesting. It presents a comprehensive literature review in this area, which will be useful for anybody embarking on their research journey in this area. Furthermore, the book provides an in-depth discussion of results obtained as a part of Lokesh's Ph.D. research.

Singapore
August 2016

Associate Professor Francis E.H. Tay

Parts of this thesis have been published in the following journal articles:

1. L. Dhakar, S. Gudla, X. Shan, Z. Wang, F.E.H. Tay, C.-H. Heng, C. Lee, Large scale triboelectric nanogenerator and self-powered pressure sensor array using low cost roll-to-roll UV embossing. Scientific Reports **6**, 22253 (2016)
2. L. Dhakar, P. Pitchappa, F.E.H. Tay, C. Lee, An intelligent skin based self-powered finger motion sensor integrated with triboelectric nanogenerator. Nano Energy **19**, 532–540 (2016)
3. L. Dhakar, F.E.H. Tay, C. Lee, Development of a broadband triboelectric energy harvester with SU-8 micropillars. J. Microelectromech. Syst. **24**(1), 91–99 (2015)
4. L. Dhakar, F.E.H. Tay, C. Lee, Investigation of contact electrification based broadband energy harvesting mechanism using elastic PDMS microstructures. J. Micromech. Microeng. **24**(10), 104002 (2014)

Acknowledgements

I would like to express my sincere gratitude to my supervisors, Profs. Francis Eng Hock Tay and Chengkuo Lee, for their continuous encouragement and unwavering support. I gratefully acknowledge Prof. Tay for his guidance throughout the journey of Ph.D. and spending time to read through the thesis. The journey of Ph.D. would never have been possible without Prof. Lee's support. He provided me with all the resources and his valuable time for all the long discussions whenever needed. He has taught me the value of hard work in any profession which has served as a moral compass through the pursuit of my Ph.D. I heartily acknowledge Prof. Andrew Wee for serving as the Chairperson of my Thesis Advisory Committee and providing valuable feedback during the Ph.D.

I would like to thank the lab staff at Centre for Integrated Circuit Failure Analysis and Reliability (CICFAR), Information Storage Materials Laboratory, Microelectronics Lab and Institute of Microelectronics. I thank Mrs. Ho for doing a great job at running the day-to-day operations at CICFAR lab, Mrs. Linn Linn for her administrative help and Mr. Koo for the technical help. I would also like to thank Dr. Zhiping Wang and Dr. Xuechuan Shan at Singapore Institute of Manufacturing Technology for their support for device fabrication and time for technical discussions.

I would like to express my gratitude to all my colleagues and friends for their support, time and energy for all kind of technical and non-technical discussions. I would like to thank Prakash, Dr. Avinash and Suvra for their company and encouragement during my time at the lab. Dr. Huicong was a great help and guide in the initial days of my research on energy harvesting devices. I would also like to thank our other group members including Zhuolin, Wang Hao, Qiongfeng, Dr. Qian You, Chong Pei and Wang Tao. The Ph.D. journey would have not been as memorable without the great friends I have made here. I thank my friends Prasanta, Abhijeet 'Patra', Prhashanna 'Ammu', Michal, Rishav and Maninder for their friendship and support.

None of this would have been possible without the love and support of my family. I cannot thank enough my Mummy and Papa, for always giving me the freedom to make my life choices and believing in me. They have been there to support and encourage me in all my endeavors. Another person I would like to thank is my sister Nisha or 'Cheena', as we affectionately call her, for always being there for me.

Contents

Acronyms

3-MPA	3-mercaptopropionic
AAO	Anodized aluminum oxide
AC	Alternating current
AFM	Atomic force microscope
Al	Aluminum
AlN	Aluminum nitride
Ar	Argon
Cu	Copper
DAQ	Data acquisition
DRIE	Deep reactive ion etching
emf	Electromotive force
FEP	Fluorinated ethylene propylene
Hg	Mercury
HMDS	Hexamethyldisilazane
IoT	Internet of Things
IPA	Isopropyl alcohol
ITO	Indium tin oxide
KOH	Potassium hydroxide
LCP	Liquid crystal polymer
LED	Light emitting diodes
LS-TENG	Large-scale triboelectric nanogenerator
MEMS	Microelectromechanical systems
NaCl	Sodium chloride
O_2	Oxygen
PA	Polyamide
PAN	Personal area network
PCB	Printed circuit board
PDMS	Polydimethylsiloxane
PECVD	Plasma enhanced chemical vapor deposition
PEH	Piezoelectric energy harvester

PET	Polyethylene terephthalate
PMMA	Polymethyl methacrylate
Pt	Platinum
PTFE	Polytetrafluoroethylene
PVDF	Polyvinylidene fluoride
PZT	Lead zirconate titanate
RFID	Radio-frequency identification
RIE	Reactive ion etching
RMS	Root mean square
SEM	Scanning electron microscope
Si	Silicon
SiN	Silicon nitride
SiO_2	Silicon dioxide
TENG	Triboelectric nanogenerators
TiO_2	Titanium dioxide
UV	Ultraviolet
ZnO	Zinc oxide

Symbols

A_{noise}	Amplitude of noise
A_{signal}	Amplitude of required signal
F_0	Amplitude of the sinusoidal excitation force
A_i	Area of the *ith* coil in electromagnetic energy harvester
S	Area of triboelectric layer
F	Average impact force
$v_{av,i}$	Average velocity of the object between P_{i-1} and P_i
$a_{av,i}$	Average velocity of the object between P_{i-1} and P_i
ΔA	Change in area
ΔP	Change in momentum
Δl	Change in the height/length
q	Charge
P	Collective dipole in material
$V_{1/2}$	Contact potential difference of metal 1 against metal 2
C_{FD}	Coupling capacitance between the device electrode and epidermis
C_{B1}	Coupling capacitance of epidermis with current carrying conductor
C_{B2}	Coupling capacitance of epidermis with ground
c	Damping coefficient
Δx	Deformation along x-axis
Δy	Deformation along y-axis
x_{PDMS}	Deformation in PDMS layer
x_{SU-8}	Deformation in SU–8 micropillar
ε_0	Dielectric constant of air
P_k	Dipole created within a molecule
y	Displacement of proof mass
ΔT	Duration of impact
k_{eff}	Effective spring constant of PDMS micropad array
k_{SU-8}	Effective spring constant of the SU–8 micropillar array
d_0	Effective thickness constant
ω	Excitation frequency

F_k	External force applied on the molecule in piezoelectric model
g	Gap between the micropillars
h	Height of the micropillar
C_O	Internal capacitance of oscilloscope
R_0	Internal resistance of oscilloscope
l	Length of the micropillar
B	Magnetic field
σ_{ii}	Normal stress in direction $i = x, y, z$
V_{oc}	Open circuit voltage
Φ	Phase difference
v	Poisson's ratio
V_A	Potential at the current carrying conductor
V_B	Potential at the device electrode
m	Proof mass
M	Proof mass at the end of cantilever
ε_{r1}	Relative dielectric constants of triboelectric layer 1
ε_{r2}	Relative dielectric constants of triboelectric layer 2
$x(t)$	Relative distance between two triboelectric layers
$v(\mathrm{t})$	Relative velocity of triboelectric layers
ω_r	Resonant frequency
I_{sc}	Short circuit current
L	Side length of the array
SNR	Signal to noise ratio
k	Spring constant
$k_{SU\text{–}8\ pillar}$	Spring constant of a single SU–8 micropillar
k_i	Spring constant of an individual PDMS micropad
k_{PDMS}	Spring constant of the PDMS layer
ε_{ii}	Strain in direction $i = x, y, z$
S	Strain in the piezoelectric layer
σ	Surface charge density due to triboelectric effect
v_n	The normal component of velocity for the top part
d_1	Thickness of triboelectric layer 1
d_2	Thickness of triboelectric layer 2
t	Time
Φ	Total magnetic flux
z	Vertical displacement of the cantilever tip
V	Voltage generated across energy harvesting device
w	Width of the micropillar
Φ_1	Work function of metal 1
Φ_2	Work function of metal 2
E	Young's modulus

List of Figures

List of Tables

Summary

Triboelectric effect has been proposed as a method to harvest mechanical energy from various sources in the environment. It also has applications in the area of sensors including motion sensors, pressure sensors and chemical sensors. This thesis explores the applications of triboelectric effect based contact electrification for mechanical energy harvesting and self-powered sensing. A comprehensive study has been conducted using micropatterned arrays to improve the understanding of device design parameters for triboelectric nanogenerators (TENG) cantilever configuration. Patterned layers were found to have better performance than unpatterned triboelectric layers. In the investigation, it was found that introduction of air voids with an optimum size was necessary to achieve better performance for TENG. It was also concluded that the performance of the TENGs could be significantly improved using smaller sized structures/patterns on the triboelectric layers. The optimum power produced by the device was measured to be 0.91 μW at 1g.

To develop wearable energy harvesting devices, human skin has been proposed and demonstrated as a potential triboelectric layer for biomechanical energy harvesting. The device generated a peak voltage of 70 V and peak current density of 2.7 $\mu A/cm^2$. The skin-based TENG has been demonstrated to capture mechanical energy from various human activities. The design for triboelectric nanogenerator has also been developed into a capacitance based sensor to accurately capture the finger movement. The triboelectric nanogenerator integrated with sensor can be used for osteoarthritis rehabilitation and advanced gesture input devices.

A novel mechanism based on triboelectric effect has been proposed to differentiate between the physical interactions in normal and sliding directions. This design can be used for tactile sensing applications. The device has also been demonstrated to harvest mechanical energy in multiple directions. This small form factor device can be used as a tactile sensor and also an energy harvester to capture energy from keyboard strokes or switches. The device was observed to generate a peak voltage and current of ~90 V and ~0.57 μA using normal force, respectively. Through the sliding motion, the peak voltage and current generated were measured to be ~5 V and ~0.08 μA, respectively.

For large-scale fabrication of triboelectric sensors and nanogenerators, a scalable fabrication process has been developed. Roll-to-roll UV embossing is used to pattern large-scale polymer films to improve the device performance. The fabricated large-scale nanogenerator produced a peak power density of 62.5 mW m^{-2}. The large-scale fabrication has been demonstrated to fabricate a large size pressure sensor array. The pressure detection sensitivity was observed to be 1.33 V kPa^{-1}. The sensor array has also been characterized and demonstrated for applications in posture monitoring, security systems and patient monitoring.

Chapter 1
Introduction

1.1 Motivation

As the human race strives to improve the quality of life, we have become increasingly surrounded by consumer electronic devices and sensors. These devices and sensors are constantly evolving into an interconnected network also termed as Internet of Things (IoT). Miniaturization of sensors and microprocessors are driving this trend forward. It is expected that by 2020, by a conservative estimate nearly 50 billion devices and sensors will be connected with each other (Fig. 1.1) [1]. In future, these devices will change the way we track the environmental conditions, air/water quality, structural health and human body condition. But as the number of devices and sensors increases, they also pose a great technical challenge of *powering* these sensors and devices in order to operate and perform necessary functions i.e. sensing, collecting, transmitting and analyzing data (Fig. 1.2). One of the main components of these electronic devices are batteries needed in order to fulfill the power requirements. As the batteries can only store limited amount of charge, it puts a restriction on the duration of time for which the devices and sensors can be powered. Another huge challenge is the disposal of batteries after the end of their lifetime. The elements used in the dry batteries such as lithium, mercury and cadmium pose severe risks for health of living beings and the environment [2].

The electronic devices are following a trend of miniaturization, driving down the production cost which is leading to further increase in the number of devices. As the number of devices increases in forms of wearable gadgets and sensors, each individual will own and use a large number of these devices. This will result in managing the power requirements and it will become practically very difficult to power the IoT completely through batteries. Therefore it becomes imperative to move onto newer, greener and sustainable ways of powering numerous electronic devices and sensors enabled by the IoT.

Harvesting energy from the surroundings is an approach to generate energy to power devices and sensors in a sustainable manner. There is enormous amount of

L. Dhakar, *Triboelectric Devices for Power Generation and Self-Powered Sensing Applications*, Springer Theses,
DOI 10.1007/978-981-10-3815-0_1

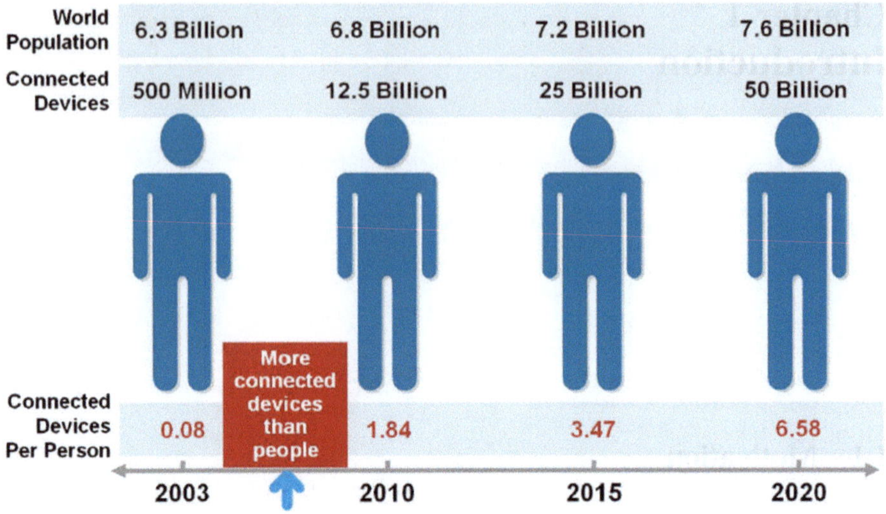

Fig. 1.1 Growth of the number of connected devices. *Source* Cisco IBSG, April 2011

Fig. 1.2 A schematic depicting component of Internet of Things (IoT). The blocks in *red color* depict the functions which require power in order to operate

energy available in our daily living environments, which is wasted without being put to any use. This energy is available in the form of motion, vibration, wind energy, solar light, heat and sound energy. Most of the energy available in these forms can be captured and used to convert to electrical energy. Sun is one of the biggest and lasting source of energy, and immense research effort has been put into developing efficient solar cells. But solar energy is subjected to many external factors e.g. weather, location and time. Therefore in order to develop energy harvesting solutions, it is important to design and develop novel solutions to harvest in other forms such as motion, sound waves and heat. The energy generated by these devices can be used in conjunction with the batteries to fulfil the power requirements of electronic devices and sensors. This will lead to the development of self-powered systems with an ability to harvest energy from their surroundings to meet their power needs. These self-powered systems will have applications in biomedical, environmental/infrastructure monitoring, security, sports training, healthcare and defense technology.

1.2 Mechanical Energy Harvesting

Mechanical energy is one of the most ubiquitous sources of energy around us. It is available in the form of human movement, breathing, heartbeat, vehicle motion and vibrations (Table 1.1). This mechanical energy is a result of conversion from one form to another e.g. chemical energy stored is converted to motion using metabolism for human activities and energy stored in fuel is converted to mechanical energy by burning the fuel in case of vehicle motion. This mechanical energy is generated through activities which are done as part of daily lives to serve a specific purpose e.g. to move from one place to another, to lift object, to pump blood to different body organs etc. Table 1.2 summarizes the mechanical power available from various human activities. During these activities, solutions can be developed to capture and scavenge the energy for useful purposes without necessitating to work specifically to generate energy. Moreover, other sources of energy including vibrations, vehicle motion and wind energy can be harvested to convert into electrical energy.

Traditionally, three principles have been used for the conversion of mechanical energy into electrical energy at relatively smaller scale:

- Piezoelectricity
- Electromagnetism
- Electrostatics

Piezoelectric mechanism is one of the most popular strategies to harvest ambient mechanical energy and is based on property of the material known as

Table 1.1 Sources of mechanical energy in the day-to-day surroundings

Human activity	Walking, running, lifting objects, breathing, heart palpitation, typing, hand movement, blood flowing through the blood vessels, speaking
Inside home	Door closing and opening, pressing switch, vibrations from household appliances e.g. microwave, air conditioner, refrigerator etc.
Outside home	Wind energy, vehicle movement on the roads, vibration from vehicles e.g. engine and chassis, rotational kinetic energy from tires
Industrial plant	Movement of vehicles, vibrations from machines, human movement

Table 1.2 Biomechanical energy/power available from the different human activities

Human activity	Mechanical power available
Blood flow	0.93 W
Exhalation	1.00 W
Upper limb movement	3.00 W
Finger typing	6.9–19 mW
Walking	67 W

piezoelectricity [3]. This property is found in materials such as lead zirconate titanate (PZT), zinc oxide (ZnO) and aluminum nitride (AlN). These materials have been used to fabricate microelectromechanical systems (MEMS) based [4] and macro scale [5] piezoelectric energy harvesting devices. In recent years, nanofabrication techniques have been used to fabricate piezoelectric energy harvesters which are more commonly referred to as nanogenerators [6, 7]. Electromagnetic energy harvesting devices work on the principle of Faraday's law of induction [8]. These devices comprise of magnetic materials and micro-fabricated coils. The micro-fabrication techniques pose design limitations of miniaturization of devices [9]. Electrostatic energy harvesters are based on a parallel plate capacitor system. These devices require external bias voltage in order to operate. Some of these devices use electrets as internal bias to convert mechanical energy into electrical energy [10].

Recently, a new principle known as triboelectric mechanism has been proposed as a new method for mechanical energy harvesting [11]. Triboelectric mechanism has considerable advantages over other mechanism described earlier. These devices have high power output, low cost, easy fabrication and wider choice of materials for fabrication. This mechanism uses contact electrification and electrostatic induction in conjunction to convert mechanical energy into electrical energy. Various micro/nanofabrication techniques including soft lithography, nanoparticle deposition, chemical/physical etching and block polymer self-assembly have been utilized for fabrication of these energy harvesting devices. The detailed working mechanism of these devices is explained in Chap. 2. Mechanical energy harvesters based on triboelectric mechanism are more commonly termed as triboelectric nanogenerators (TENGs).

1.3 Scope and Organization of Thesis

This thesis is aimed at developing mechanical energy harvesting solutions using triboelectric mechanism. The overarching objective of energy harvesting technology is to realize self-powered sensor systems and electronic devices. The research work conducted for this thesis utilizes triboelectric effect to develop self-powered sensor systems, which can function without the need of external power. These self-powered systems will have immense applications in future in the area of healthcare, security systems, remote patient monitoring and rehabilitation. If any technology has to be taken to masses and commercialized, it is important to make the technology scalable to take advantage of economies of scale leading to low cost of the devices. The research conducted also demonstrates the scalable fabrication of triboelectric energy harvesting devices and sensors. The organization of the thesis as per research work conducted is explained hereafter.

To start with, a comprehensive literature review was conducted. Chapter 2 details a review of mechanisms used for mechanical energy harvesting including piezoelectric, electromagnetic, electrostatic and triboelectric. As the main focus of this thesis is using the triboelectric effect to design and develop mechanical energy harvesters, special attention has been paid to the triboelectric mechanism and TENGs. The working principle of the TENGs has been explained. Thereafter fabrication techniques used to realize TENGs have been reviewed. The next part of the literature review was conducted for the two major directions of the research: (1) triboelectric mechanism based energy harvesting devices and (2) triboelectric mechanism based self-powered sensors.

In order to conduct design and develop mechanical energy harvesting devices, an understanding of the working mechanism and device operation is necessary. Keeping this objective in mind, a detailed study was conducted to study the effect of surface topography on the TENG device performance as described in Chap. 3. The study was carried out using micro fabricated arrays of SU-8 micropillars and elastic PDMS micropad structures. The device design used for the experiments is shown in Fig. 1.3a. Experiments were designed by changing the microstructure dimension

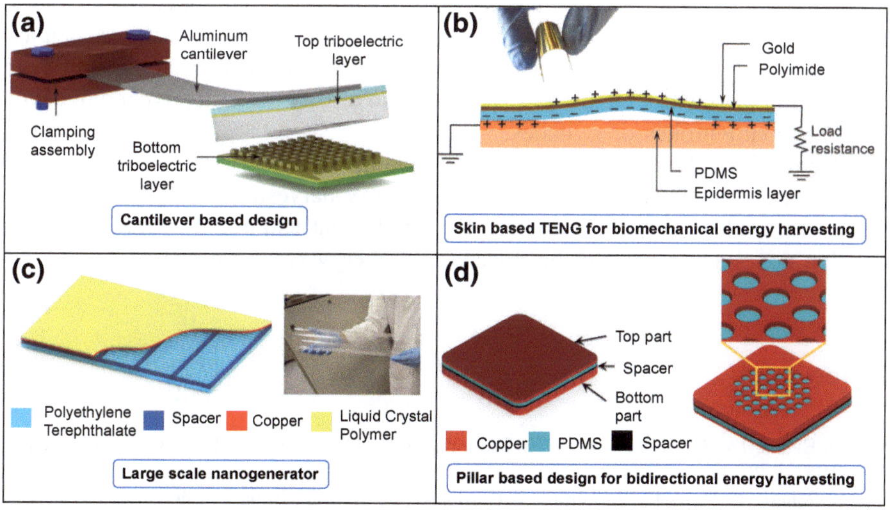

Fig. 1.3 Summary of the proposed device designs in the thesis. **a** Cantilever based design used to study the effect of topography on the performance of TENG. **b** Skin based wearable TENG which can be used to harvest biomechanical energy harvesting. The harvested energy can possibly be used to power small wearable sensors and/or other devices. **c** Schematic of large scale nanogenerator design. The photograph shows the patterned PET film fabricated using roll-to-roll UV embossing process. **d** Schematic of the pillar based design. This design can be used to harvest energy from both normal and shear force

and microstructure area density for different array configurations. This variation of microstructure array was used to systematically study the effect of microstructure array configuration on the output performance of TENG.

Chapter 4 describes the development of TENGs for wearable energy harvesting applications. This chapter explores human skin (epidermis) as a triboelectric material for wearable energy harvesting applications. The schematic of the proposed design is shown in Fig. 1.3b. The chapter explains the fabrication of flexible devices with micro/nano structures to improve the performance of wearable skin based TENGs. These skin based TENGs were developed to harvest energy from muscle movement from day-to-day activities. These devices were also developed as self-powered sensors for applications in human motion tracking, advanced gesture devices, human machine interfacing and rehabilitation.

One of the major bottlenecks in bringing the TENG devices closer to market and consumers is lack of scalable fabrication methods for large scale and low cost manufacturing of these devices. Chapter 5 proposes the fabrication of TENG and triboelectric mechanism based self-powered sensors using scalable fabrication methods. Roll-to-roll ultraviolet (UV) embossing was used to pattern large polymer films, which were used to fabricate large size TENG as shown in Fig. 1.3c. Self-powered pressure sensor array was also designed and demonstrated which can be fabricated at very low cost using large scale processing techniques. The fabrication process flow was proposed and developed which can be used for industrial scale fabrication of TENG and self-powered sensor devices.

Chapter 6 proposes a novel a mechanism design to differentiate between the mechanical interactions in normal and sliding directions. The device uses PDMS micropillars inserted in the etched holes as shown in Fig. 1.3d. The device can also be used to harvest mechanical energy in multiple directions. This small form factor device can be used as a tactile sensor and also an energy harvester to capture energy from keyboard strokes or switches.

Chapter 7 concludes the thesis with key findings and developments in the area of triboelectric mechanism based mechanical energy harvesting. It also proposes the future work which will be conducted to complement and continue the research to develop novel and more efficient mechanical energy harvesting devices.

References

1. D. Evans, *The Internet of Things*, in Presented at the Cisco Internet Business Solutions Group (IBSG), 2011
2. S. Panero, C. Romoli, M. Achilli, E. Cardarelli, B. Scrosati, Impact of household batteries in landfills. J. Power Sour. **57**, 9–12 (1995)
3. T. Ikeda, *Fundamentals of Piezoelectricity* (Oxford University Press, Oxford, 1996)
4. H.-B. Fang, J.-Q. Liu, Z.-Y. Xu, L. Dong, L. Wang, D. Chen et al., Fabrication and performance of MEMS-based piezoelectric power generator for vibration energy harvesting. Microelectron. J. **37**, 1280–1284 (2006)

5. A. Erturk, D.J. Inman, An experimentally validated bimorph cantilever model for piezoelectric energy harvesting from base excitations. Smart Mater. Struct. **18**, 025009 (2009)
6. Y. Qin, X. Wang, Z.L. Wang, Microfibre–nanowire hybrid structure for energy scavenging. Nature **451**, 809–813 (2008)
7. S. Xu, Y. Qin, C. Xu, Y. Wei, R. Yang, Z.L. Wang, Self-powered nanowire devices. Nat. Nanotechnol. **5**, 366–373 (2010)
8. S.P. Beeby, R. Torah, M. Tudor, P. Glynne-Jones, T. O'Donnell, C. Saha et al., A micro electromagnetic generator for vibration energy harvesting. J. Micromech. Microeng. **17**, 1257 (2007)
9. S.P. Beeby, T. O'Donnell, in *Electromagnetic Energy Harvesting*, ed. Energy Harvesting Technologies (Springer, Berlin, 2009), pp. 129–161
10. E. Halvorsen, E. Westby, S. Husa, A. Vogl, N. Østbø, V. Leonov et al., An electrostatic energy harvester with electret bias. Transducers **2009**, 21–25 (2009)
11. F.-R. Fan, Z.-Q. Tian, Z.L. Wang, Flexible triboelectric generator. Nano Energy **1**, 328–334 (2012)

References

[illegible] Q. [illegible] A. [illegible] internally [illegible] and [illegible] [illegible] [illegible] [illegible] [illegible] Nano [illegible] (2009)
6. [illegible] Q. [illegible] [illegible] Wang, [illegible] [illegible] [illegible] [illegible] Nano [illegible] (2008)
7. [illegible] Y. [illegible] Y. Wei, [illegible] Wang, Z.L. Wang [illegible] [illegible] (2010)
8. S. Sheehy, R. Toroh, M. Tudor, R. Gomme-Jones, T. O'Driscoll [illegible] [illegible] (2007)
9. [illegible] [illegible] [illegible] (2014)
10. [illegible] [illegible] [illegible] Z. [illegible] Wang [illegible] (2015)
11. [illegible] [illegible] Z.L. Wang [illegible] [illegible]

Chapter 2
Overview of Energy Harvesting Technologies

Harvesting or scavenging mechanical energy from the surroundings is a potential strategy to develop self-powered sensor nodes and electronic devices. In this chapter, a literature review is presented for various mechanical energy harvesting technologies to understand the state-of-the-art. The first part of the chapter discusses commonly used principles to convert mechanical energy into electrical energy. Basic physics and working all the mechanisms have been explained and previously developed devices by other research groups have been reviewed.

The second part of the chapter reviews the progress in the area of TENGs. Different working mechanisms, materials and fabrication techniques utilized for TENG devices have been reviewed. Thereafter, applications of TENG devices to harvest mechanical energy in different forms e.g. biomechanical energy, vibration energy and wind energy, has been discussed. Triboelectric mechanism also has immense applications in the area of sensing. Triboelectric mechanism has been reviewed for various type of self-powered sensors e.g. pressure sensors, vibration sensors, motion tracking systems, chemical sensors etc.

2.1 Mechanical Energy Harvesting Mechanisms

2.1.1 Piezoelectric Energy Harvesters

Piezoelectric mechanism is the most common and well researched technique in the area of mechanical energy harvesters. These devices work on the property of material known as piezoelectricity. It is the property of the materials to produce charge (or voltage) when stimulated by a mechanical stress. Some of the popular materials which exhibit piezoelectricity are quartz, lead zirconate titanate (PZT), aluminum nitride (AlN), zinc oxide (ZnO) and polyvinylidene fluoride (PVDF). Piezoelectric effect can be explained using a molecular model as shown in Fig. 2.1.

L. Dhakar, *Triboelectric Devices for Power Generation and Self-Powered Sensing Applications*, Springer Theses,
DOI 10.1007/978-981-10-3815-0_2

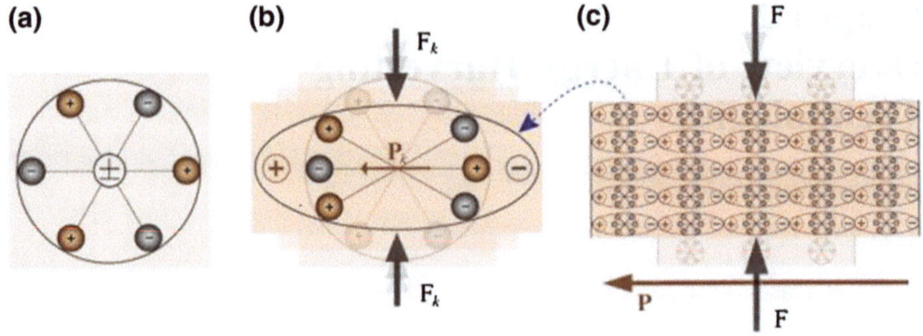

Fig. 2.1 Piezoelectricity explained using a molecular model. **a** A molecule in an undisturbed state without any external force. **b** The molecule becomes polarized due to an external force F_k applied. **c** Molecules in a piezoelectric material collectively leading to generation of voltage due to polarization when an external force is applied [1]

In Fig. 2.1a, the molecule is in undisturbed state, but as an external force F_k is applied, it causes unbalance of charge leading to creation of a small dipole P_k as shown in Fig. 2.1b. Collectively, all these small dipoles create a dipole P in the material. This phenomenon in materials is known as piezoelectricity.

Researchers have utilized the piezoelectric effect to harvest mechanical energy and convert it into electrical energy. The most common configuration and design used for piezoelectric energy harvester (PEH) is using a cantilever [2, 3]. A typical configuration of cantilever based PEH is shown in Fig. 2.2. The piezoelectric cantilever undergoes cycle of compression and tension as it is excited through external vibration or cyclic force. This cyclic force results into generation of cyclic voltage signal across the cantilever electrodes across the piezoelectric layer. A proof mass (*M*) at the tip of the cantilever beam is typically used to improve the stress levels in the piezoelectric layer. This results in a higher magnitude of the voltage generated at the electrodes.

Renaud et al. [4] used a microelectromechanical system (MEMS) based PZT cantilever design to harvest vibrational energy. The major limitation of cantilever

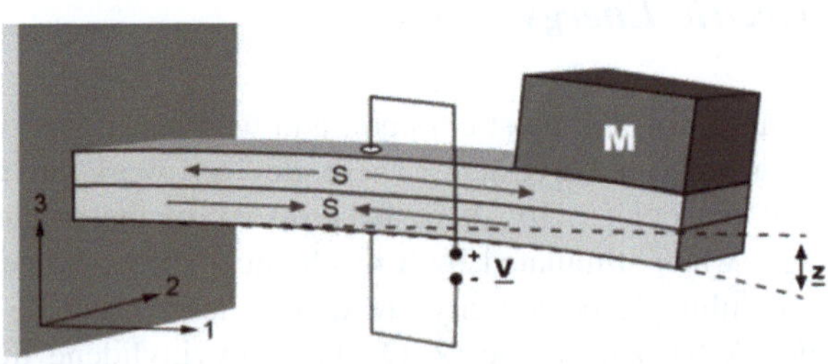

Fig. 2.2 Schematic of a typical cantilever based PEH. *S* is the strain in the piezoelectric layer shown with directional arrows, *V* is the generated voltage, *M* is the proof mass and *z* is vertical displacement of the cantilever tip [2]

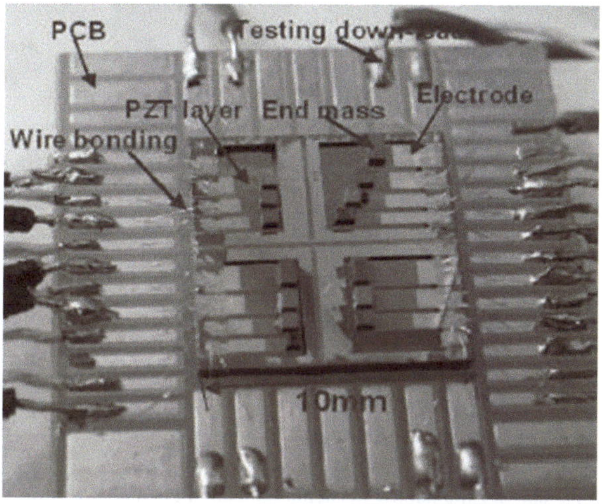

Fig. 2.3 Array of cantilevers used to improve the operating bandwidth of the vibrational PEH [5]

based PEH is the limited frequency range in which these devices can operate due to resonant behavior. Liu et al. [5] used an array of PZT cantilevers (Fig. 2.3) with different mechanical dimensions and hence different resonant frequencies to improve the operating bandwidth of the devices.

Apart from the cantilever based designs of PEH devices, researchers have used piezoelectric materials for demonstrations in many practical applications. Kymissis et al. [6] from MIT Media Laboratory first demonstrated shoe based energy harvesting system in 1998 using PZT and PVDF for wearable applications. The image of the prototype device is shown in Fig. 2.4. The prototype was demonstrated to generate enough power to for a radio-frequency identification (RFID) tag. A windmill based piezoelectric energy harvesting was developed by Priya [7] using 10 piezoelectric bimorphs which were actuated using a windmill.

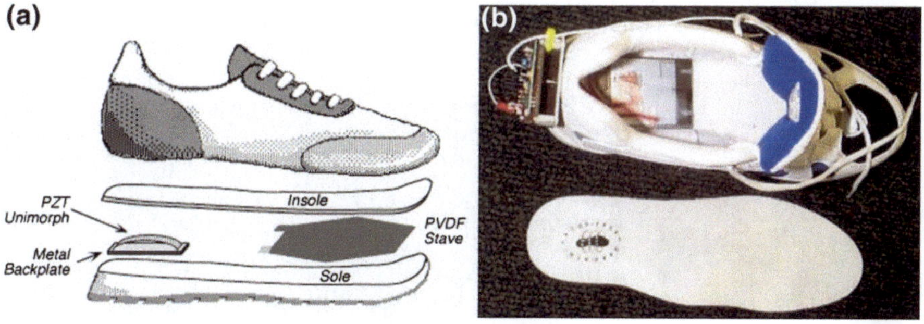

Fig. 2.4 **a** Schematic diagram of the shoe based energy harvesting device. **b** Photograph of the fabricated prototype [6]

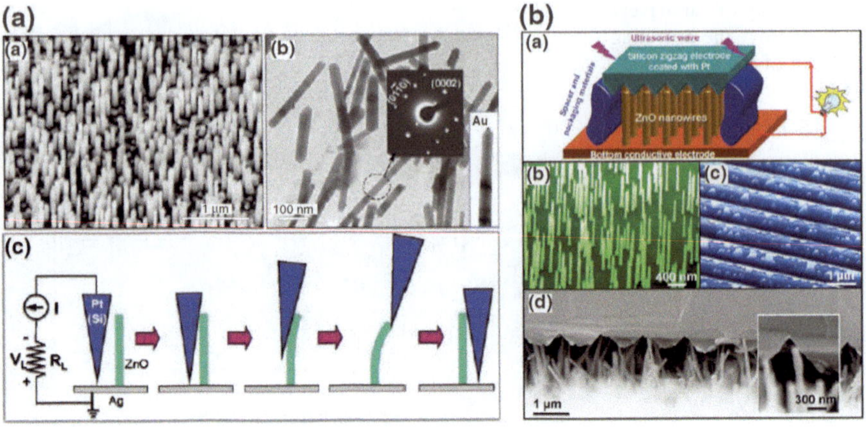

Fig. 2.5 **a** ZnO nanowires actuated using Pt coated AFM tip [8]. **b** Pt coated AFM tip replaced by a zigzag Pt electrode actuated using ultrasonic waves [9]

ZnO nanowire based energy harvester was first demonstrated by Wang and Song [8] using array of aligned nanowires (Fig. 2.5a). The nanowires were deflected with a conductive atomic force microscope (AFM) tip in contact mode. To improve the practical application of the concept, the Pt coated AFM tips were replaced by microfabricated zigzag shaped Pt electrodes as shown in Fig. 2.5b [9]. Chung et al. [10] fabricated ZnO thin film on a polyethylene terephthalate (PET) substrate using simple reactive zinc hydroxo-condensation process (Fig. 2.6a). This device was demonstrated to be suited for human wearable energy harvesting devices. Lee et al. [11] fabricated a flexible and conformable ZnO nanowire based piezoelectric nanogenerator in an aluminum (Al) foil coated with Polymethyl methacrylate (PMMA) (Fig. 2.6a). The device could generate an output voltage and current of up to 50 mV and 200 nA respectively, by utilizing fluttering motion generated through wind energy. Lin et al. [12] demonstrated a transparent flexible nanogenerator by growing ZnO nanowires on a PDMS substrate. The generated an output voltage and

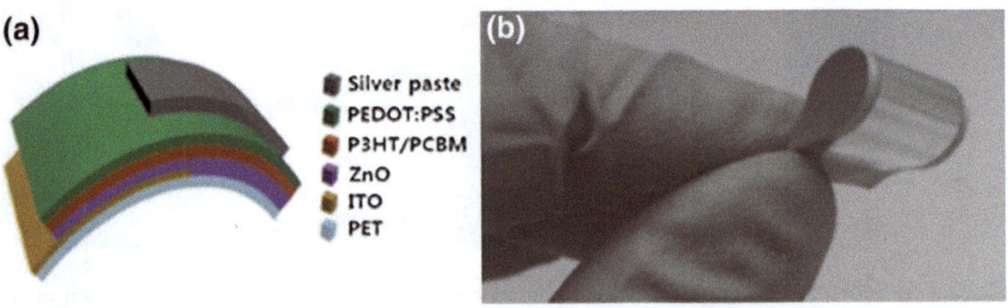

Fig. 2.6 **a** Schematic diagram of the ZnO film based piezoelectric nanogenerator on PET substrate [10]. **b** Photograph of the super flexible piezoelectric nanogenerator [11]

current of 8 V and 0.5 μA, respectively, with a power output density of 5.3 mW/cm^3. The device was also demonstrated as a self-powered sensor to monitor vehicle speed and detect vehicle weight.

Research breakthroughs in the area of novel fabrication techniques for piezoelectric materials have found applications in the area of wearable and flexible energy harvesting devices. But limited choice of materials and complex fabrication techniques still remain one of important challenges for piezoelectric energy harvesting devices. These fabrication challenges have been a bottleneck to find application implantable, biocompatible and wearable energy harvesters, which can be commercialized.

2.1.2 Electromagnetic Energy Harvesters

Electromagnetic mechanical energy harvesters work on the principle of Faraday's law of induction. Majority of the electrical motors, transformers, inductors and generators are based on this fundamental operating principle. Faraday's law states that the voltage or electromotive force (emf) generated in a closed circuit directly proportional to the total flux through the closed loop circuit. The equation for Faraday's law is given by:

$$V = -\frac{d\Phi}{dt} \tag{2.1}$$

where V the voltage is induced across the two ends of closed circuit, Φ is total flux and t is time. If there are total N coils in the circuit, total flux Φ is given by:

$$\Phi = \sum_{i=1}^{N} \int_{A_i} B.dA \tag{2.2}$$

where B is the magnetic field and A_i is the area of the ith coil. Faraday's law has been extensively used in large sized commercial generators to produce power using fuel, hydropower and other traditional sources. Same principle can be utilized to harvest mechanical energy at a much smaller scale. Williams et al. [13] demonstrated one of the first micro scale electromagnetic energy harvester using a samarium-cobalt permanent magnet and planar gold (Au) coils. Lee et al. [14] demonstrated a AA battery sized electromagnetic energy harvester using laser micromachining as a fabrication technique (Fig. 2.7a). Same research group also demonstrated an electromagnetic energy harvester capable of harvesting energy from different resonant modes as shown in Fig. 2.7b.

Kulkarni et al. [18] developed a micro energy harvester by using electrodeposited copper coils and NdFeB magnetics. The device had a small volume of 0.1 cm^3 and could generated a power of 586 nW across a load resistance of 110 Ω at an acceleration of 8.829 m/s^2. Cepnik and Wallrabe [16] designed a flat

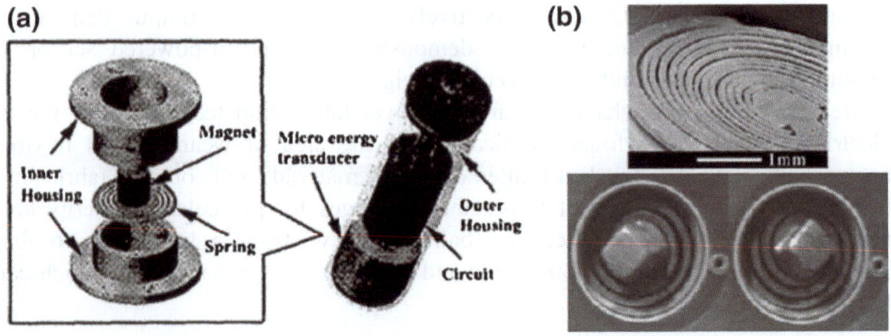

Fig. 2.7 **a** AA size micro energy harvester by Lee et al. [14]. **b** Multi-modal energy harvester developed by Chin et al. [15]

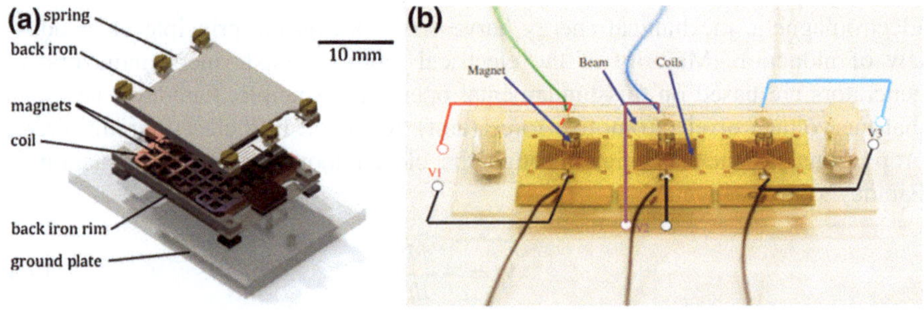

Fig. 2.8 **a** A flat energy harvester using serpentine springs for in-plane movement [16]. **b** A low cost electromagnetic energy harvester using coils etched on a printed circuit board (PCB) [17]

electromagnetic energy harvester which used micro fabricated serpentine springs for in-plane movement. The device as shown in Fig. 2.8a generated an average power of 12 μW at an acceleration of 1 ms^{-2}. Yang et al. [17] used an acrylic fixed-fixed beam with three permanent magnets attached to it. Correspondingly, three coils were fabricated on a printed a printed circuit board (PCB) and placed beneath the permanent magnets as shown in Fig. 2.8b. As the beam was excited using vibrations, it generated power from different vibration modes of the beam. The maximum power produced by device was demonstrated to be 3.2 μW.

Donelan et al. [19] demonstrated a complete working prototype of a biomechanical energy harvester using electromagnetic energy harvesting technique. The device was designed for applications for powered prosthetic limbs and charging medical devices (Fig. 2.9). The electromagnetic mechanism based energy harvesters have advantage of very low device impedance as compared to piezoelectric, electrostatic and triboelectric energy harvesting devices. These devices are suitable for large scale energy production but as the size of the device decreases, the coil and magnet fabrication becomes a huge challenge.

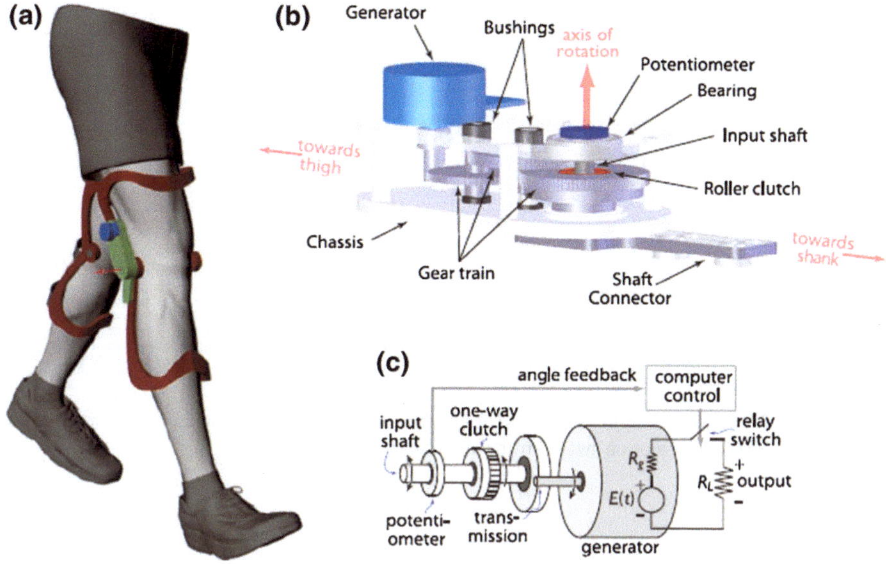

Fig. 2.9 Biomechanical energy harvester to harvest energy during human walking for applications in prosthetic limbs [19]

2.1.3 Electrostatic Energy Harvesters

Electrostatic energy harvesters work on the principle of variable parallel plate capacitor. The mechanical energy available in the surroundings is used to change the gap between two parallel plates, which together form the capacitor system. This change in the gap or relative position of the two plates leads to flow of charges in the external load connected across the two plates. The physical mechanism of the electrostatic energy harvesters can be distinguished based on the type of relative motion between the two plates as shown in Fig. 2.10.

Electrostatic energy harvesters can be divided into categories: (i) using external voltage supply for biasing (ii) using electrets. The first category of the devices rely on an external bias to create a potential difference between the two parallel plates. In the second type of device, electrets are used to create a potential difference between the parallel plates. Electrets are essentially dielectric materials with permanent electrical polarization analogous to permanent magnet.

Meninger et al. [21] proposed one of the first energy harvesters based on the electrostatic mechanism. To improve the overall capacitance per unit displacement of the moving part, Hoffmann et al. [22] used triangular electrode design as shown in Fig. 2.11a. Yang et al. [23] demonstrated a rotary comb based energy harvester design for planar vibrations. The device produced a maximum power of 2.5 μW at an acceleration of 2.5 g. Nguyen et al. [24], Nguyen and Halvorsen [25] proposed and demonstrated the concept of utilizing non-linear springs as shown in Fig. 2.12b

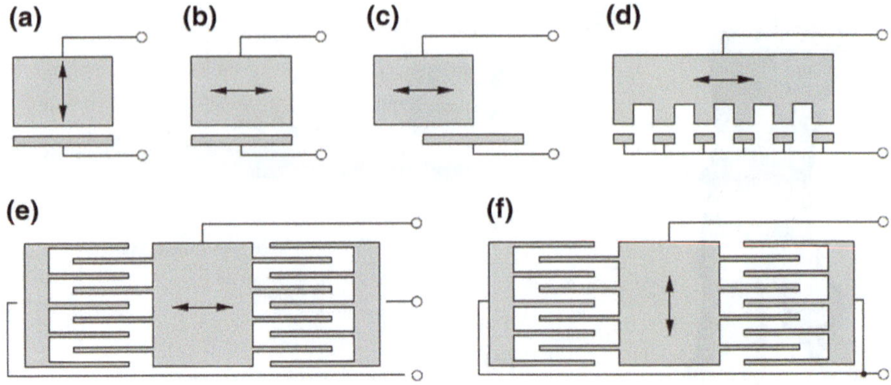

Fig. 2.10 **a–f** Various configurations used for electrostatic energy harvesters. The *arrows* depict the direction of relative motion between the fixed and moving part [20]

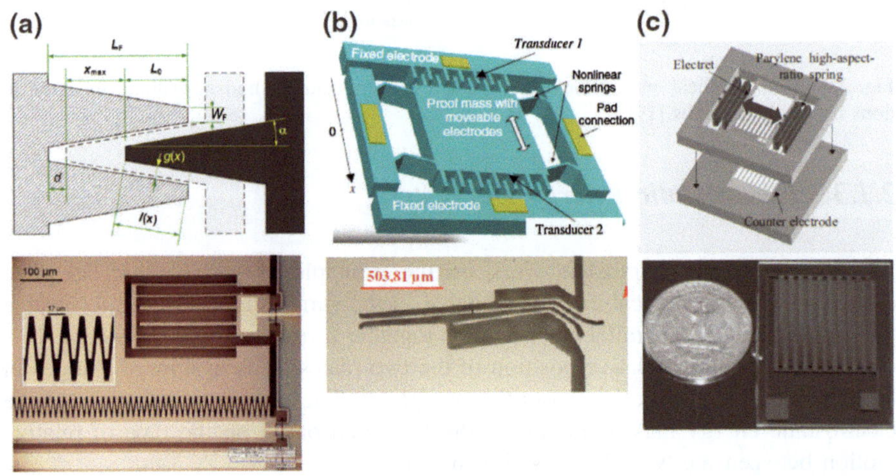

Fig. 2.11 **a** Triangular electrodes used in the electrostatic energy harvesters to increase the capacitance change per unit of displacement [22]. **b** Non-linear springs to improve the operating bandwidth of vibration based electrostatic energy harvesters [24]. **c** Vibrational energy harvester using CYTOP electret [27]

to improve the operating frequency bandwidth of the device. The non-linear springs were designed to exhibit softening in leading to wideband characteristics and higher output power.

Boland et al. [26] proposed a micro fabricated Teflon electret based energy harvester utilizing rotation mechanism. The rotational electrostatic energy harvester generated a power of 25 μW. Tsutsumino et al. [27] used a CYTOP based electret to harvest vibrational energy which could produce a power of up to 38 μW (Fig. 2.11c). They later improved the surface charge density of the electrodes by the

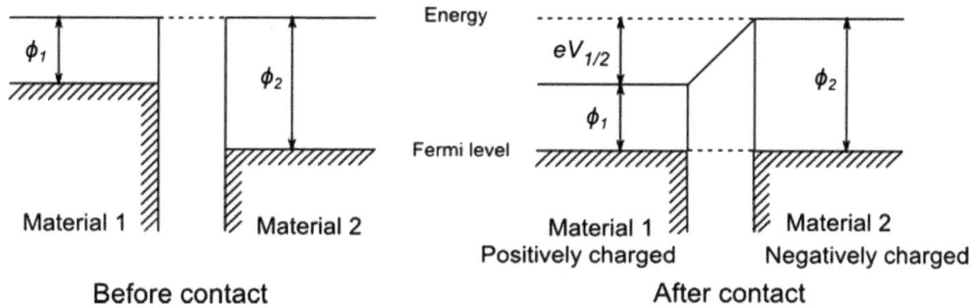

Fig. 2.12 Charge transfer due to the difference in work function between two metal for metal-metal contact [29]

addition of amidosilyl end group [28]. The improved device generated power output up to 0.7 mW.

The electrostatic energy harvesters are not the most popular solution for mechanical energy harvesting because of their low power output as compared to other mechanisms. Moreover, electret less electrostatic energy harvesters need an external voltage supply which is not suitable for self-powered wireless nodes. On the other hand, electret based electrostatic energy harvesters retain the charges only for a limited period of time.

2.2 Triboelectric Energy Harvesting

Triboelectric energy harvesters utilize two principles working in conjunction: triboelectric effect and electrostatic induction. Triboelectric effect is a type of contact electrification which leads to charging of two surfaces when there are put in contact with each other. Many theories have been proposed in literature to explain the phenomenon of contact electrification [29]. One of the well-established and accepted model to explain the concept of electron transfer is based on difference in work functions of various materials [30]. As per this model, as two metals come in contact with each other, the electrons transfer by tunneling in order to prevail the thermodynamic equilibrium. The potential difference $V_{1/2}$ generated between the two metals due to this charge transfer can be given by the following equation:

$$V_{1/2} = -\frac{(\phi_1 - \phi_2)}{e} \tag{2.3}$$

where $V_{1/2}$ is the contact potential difference of metal 1 against metal 2, ϕ_1 is the work function of metal 1, ϕ_2 is the work function of metal 2 and e is the elementary charge. The same theoretical model can be used to explain and understand the charge transfer in metal-insulator and insulator-insulator contact by using the

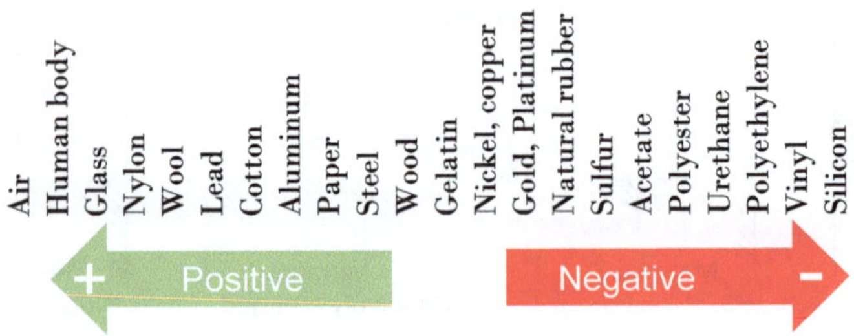

Fig. 2.13 Triboelectric series for various materials arranged in order of their tendency to attract or donate electrons

concept of effective work function for insulators [31, 32]. Figure 2.12 pictorially depicts this model given by Harper [30].

The charges generated during the contact electrification process are generated on the surface of the both the materials or triboelectric layers. Based on chemical structure and properties, various metals and dielectric materials have different tendencies to donate and accept electrons. Different metals and dielectric materials can be arranged in an order as per their tendency to donate or accept electrons. This order is known as triboelectric series [33, 34] and is shown in Fig. 2.13 for a limited number of materials. The materials on the positive side of the series have a tendency to donate electrons relative to the materials on negative side. This means that the materials on the positive side have propensity to get positive charged and vice versa. Relative positions of materials in triboelectric series can be used to predict and choose the polarity of charges generated when two material are put in contact with each other. Also farther the two triboelectric materials in series, higher is the surface charge density produced during the contact electrification process. This rule is important for selection of materials during the design of TENGs.

On the basis of mechanical operating principle, triboelectric mechanism for energy harvesting can broadly be divided into two categories: (i) Out-of-plane contact-separation mechanism (ii) in-plane sliding mechanism. The theory and working principle for both the mechanisms are explained in following sections.

2.2.1 Out-of-Plane Contact-Separation Mechanism

In out-of-plane mechanism the triboelectric layers undergo relative motion in the direction normal to the plane of contact. This relative motion results in contact and separation of the triboelectric layers. As the triboelectric layers come into contact with each other, both layers get charged with equal and opposite surface charge density σ due to contact electrification. The surface charges lead to induced charges on the metal electrodes in the energy harvesting device as shown in Fig. 2.14 [35]. As the two layers undergo relative motion i.e. separating or approaching due to

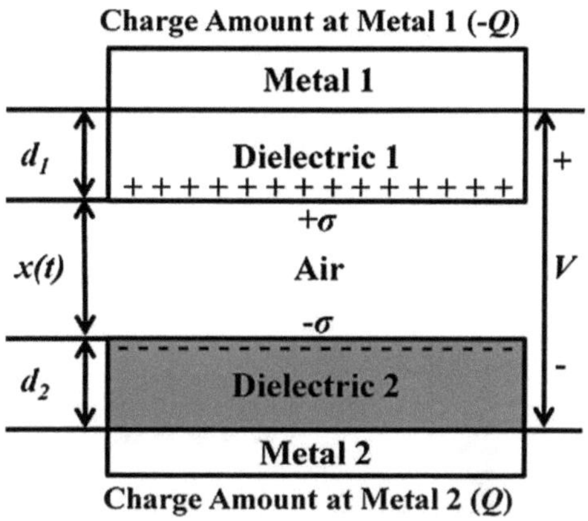

Fig. 2.14 Schematic of an out-of-plane contact-separation based triboelectric energy harvester [35]

external mechanical pressure/force, the relative potential between the two electrode changes. This change in relative potential leads to a current in the external load connected between the two electrodes. This system can be treated as a system of parallel plate capacitors and can be used to develop a theoretical model. In this case, working principle is essentially similar to the electrostatic energy harvester but the charges are internally generated using contact electrification instead of an externally power supply or using electrets. A general theoretical model for this mechanism was developed by Niu et al. [35]. As per the model, the open circuit voltage and short circuit current can be given by Eqs. (2.4) and (2.5), respectively.

$$V_{oc} = \frac{\sigma x(t)}{\varepsilon_0} \tag{2.4}$$

$$I_{SC} = \frac{S\sigma d_0 v(t)}{(d_0 + x(t))^2} \tag{2.5}$$

where V_{oc} is the open circuit voltage, σ is the surface charge density, $x(t)$ is the relative distance between two triboelectric layers, ε_0 is the dielectric constant of air, I_{SC} is the short circuit current, S is the area of triboelectric layer, $v(t)$ is the relative velocity of triboelectric layers, d_0 is defined as the effective thickness constant given by $\frac{d_1}{\varepsilon r_1} + \frac{d_2}{\varepsilon r_2}$. Here d_1 and d_2 are thickness of triboelectric layers 1 and 2, ε_{r1} and ε_{r2} are relative dielectric constants of triboelectric layers 1 and 2.

2.2.2 *In-Plane Sliding Mechanism*

As the name suggests, in-plane sliding mechanisms comprises of relative sliding motion between triboelectric layers. A schematic for in-plane sliding mechanism is shown in Fig. 2.15. As the two triboelectric layers are in contact there is an equal

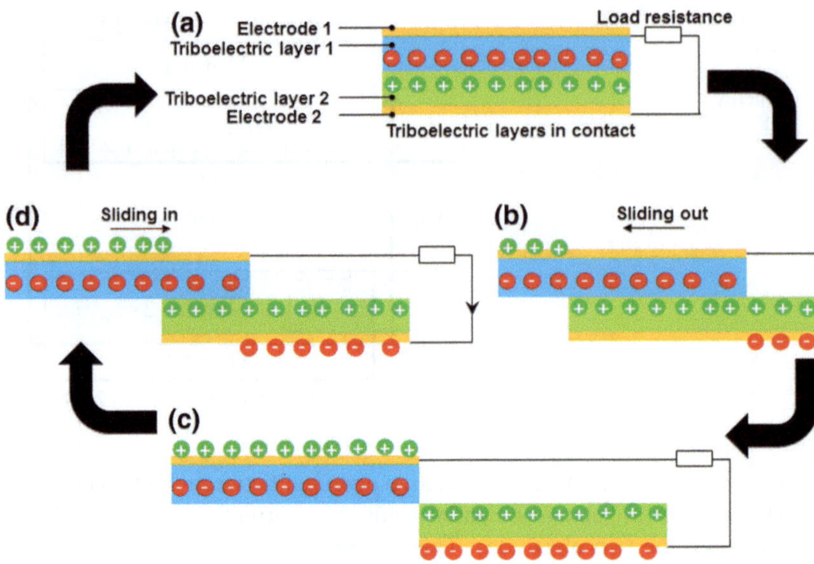

Fig. 2.15 Schematic for in-plane sliding mechanism based triboelectric energy harvester

and opposite surface charge generated on surface of both the layers. As triboelectric layer 1 moves out relative to triboelectric layer 2, there is a decrease in the contact area between two layers. This leads to charges induced in electrodes corresponding to balance the surface charges on both triboelectric layers. As both the electrodes are connected through a load resistor, there is a flow of electrons from top electrode to bottom electrode to balance the charges. This flow of electrons in the load resistor or current flow continues till the two triboelectric layers are completely separated as shown in Fig. 2.15c.

As the energy generation cycle continues, triboelectric layer 1 moves in relative to triboelectric layer 2 due to the external mechanical force. This leads to a flow of current in the external circuit in the opposite direction of sliding out process as shown in Fig. 2.15d. A theoretical model was proposed by Niu et al. [36] for the in-plane sliding mechanism based triboelectric energy harvester. According to the model, the open circuit voltage and short circuit current can be expressed by Eqs. (2.6) and (2.7).

$$V_{oc} = \frac{\sigma x}{\varepsilon_0 (l - x)} \left(\frac{d_1}{\varepsilon r_1} + \frac{d_2}{\varepsilon r_2} \right) \tag{2.6}$$

$$I_{SC} = \sigma w v(t) \tag{2.7}$$

where V_{oc} is the open circuit voltage, σ is the surface charge density, x is the relative separation distance between triboelectric layers in lateral direction, l is the length of triboelectric layers in the direction of motion, ε_0 is the dielectric constant of air, d_1 and d_2 are thickness of triboelectric layers 1 and 2, ε_{r1} and ε_{r2} are the relative dielectric constants of triboelectric layers 1 and 2, I_{sc} is the short circuit current, $v(t)$ is the relative velocity of triboelectric layers, w is the width in the lateral direction perpendicular to motion direction of the triboelectric layer.

2.3 Materials and Fabrication of Triboelectric Nanogenerators

During the design of TENG devices, contact electrification process due to triboelectric effect can be engineered to improve the power generation performance. The performance of contact electrification majorly depends on two factors: (a) choice of triboelectric materials and (b) surface topography of the triboelectric layers. Choice of materials for triboelectric layers is important as it fundamentally affects the amount of surface charge generated due to triboelectric effect. The generated surface charge depends on the relative tendency of two triboelectric materials. Other practical factors affecting the choice of materials are cost, type of application, biocompatibility and fabrication limitations. In addition, surface topography is another crucial factor as it is predominantly responsible for defining effective contact area and charge separation during contact electrification due to triboelectric effect.

Researchers have used various materials as triboelectric layers based on the fabrication compatibility, cost and patternability. Another design step to improve the performance of TENGs is to deploy different fabrication techniques to achieve various type of micro/nano surface topographies to enhance the surface charge density during contact electrification. A summary is presented in Table 2.1 about these materials and topographies used to improve the device performance.

2.4 Triboelectric Energy Harvesters and Self-powered Sensors

One of the early prototype of TENG was reported in 2012 by Fan et al. [53] based on triboelectric mechanism as shown in Fig. 2.16a. A transparent and flexible energy harvester was demonstrated which used patterned PDMS films as triboelectric layer. Three patterns including line, cube and pyramid shaped structures were used to pattern PDMS films. It was demonstrated that plain film, line, cube and pyramid patterns based device had performance in the increasing order as depicted in Fig. 2.16b. The device generated a maximum output voltage and current of 18 V and 0.7 μA, respectively. The device was also demonstrated as a

Table 2.1 Materials and surface topographies for triboelectric energy harvesters

Material	Surface topography/processing	Reference
TiO_2	Nanowires and nanosheet	Lin et al. [37]
Al	Nanopores	Yang et al. [38]
PTFE	Nanowires	Yang et al. [38]
SiO_2	Nanoparticles	Cheng et al. [39]
PDMS	Pyramid shaped structures using molding	Han et al. [40]
Al	Nanostructures of average size ~200 nm	Han et al. [40]
PDMS	Nanowire	Yang et al. [41]
Si	Pyramid shaped texture	Yang et al. [41]
PTFE	Etched nanoparticle topography	Yang et al. [42]
Ag	Nanowire/nanoparticle composite	Lin et al. [43]
Fluorinated ethylene-propylene	Dry etched nanowire structures	Yang et al. [44]
Kapton	Dispersed with PTFE nanoparticles	Jing et al. [45]
PDMS	Micropillars	Yang et al. [46]
Al	Nanopores	Bai et al. [47]
Poly-vinylchloride	Nanowires	Du et al. [48]
Cu	Micro roughness by coating Cu on etched Si structures	Guo et al. [49]
Kapton	Dry etched nanowires	Zhu et al. [50]
PDMS	Convex dome shaped structures	Hou et al. [51]
Au	Nanoparticles	Zhu et al. [52]

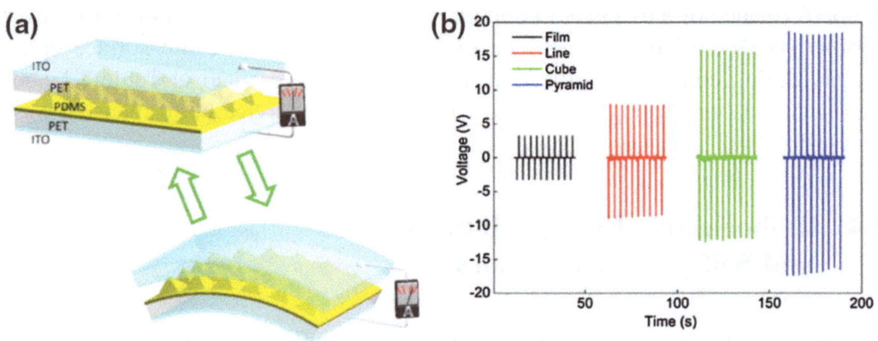

Fig. 2.16 **a** Schematic drawing of one of the early prototypes of triboelectric mechanism based energy harvester. **b** Voltage generated by the energy harvester using different topographies for patterned PDMS films [53]

self-powered pressure sensor which means the sensors do not need any external power source in order to operate as is required in the case of capacitive and piezoresistive sensors. In this Sect. 2.4, TENG devices have been reviewed based on different applications.

2.4.1 Biomechanical Energy Harvesters

Biomechanical energy is one of the most prevalent form of energy around human beings which can be used to power portable electronic devices like smartphones, watches and wearable sensors. Researchers have demonstrated various concepts and prototype devices to harvest biomechanical energy. Yang et al. [38] used an integrated rhombic gridded structure to harvest mechanical energy from vibrations generated from human walking (Fig. 2.17a). Rhombus shaped stacked structure used the hybridization of both contact-separation and sliding mode to enable contact electrification process. The TENG device was assembled on a backpack to develop a mobile source power for disaster relief workers, explorers in remote locations and field engineers. To harvest the energy directly from human heel strike, a low cost triboelectric energy harvester was developed by Bai et al. [47]. The device utilized multilayered zigzag structure in order to increase the effective contact area without significantly affecting the overall form factor of device (Fig. 2.17b). The device was assembled at the bottom of a common shoe and generated enough power to light up a few LEDs using the human walking motion. The energy harvesting device generated a maximum open circuit voltage of 215 V and a short circuit current of 0.66 mA. The maximum power density was reported to be 9.76 mW/cm^2.

Majority of the mechanical energy harvesting devices focus on harvesting energy from the external body movements like walking, hand movement etc. This energy can be used to power wearable sensors and small electronic devices. But as the technology advances, there will be many implanted sensors and devices being designed aimed at curing diseases, remote monitoring and aiding organs in functioning. Zheng et al. [54] proposed a breathing driven TENG to power in vivo devices such as pacemaker as shown in Fig. 2.18a. The device utilized micro pyramid patterned PDMS and aluminum metal layer as two triboelectric layers for energy generation. The device was also demonstrated for in vivo applications by implanting it on a living rat's diaphragm as shown in Fig. 2.18b. The implanted TENG generated a maximum power density of 8.44 mW m^{-2}, which was shown to drive a pacemaker device.

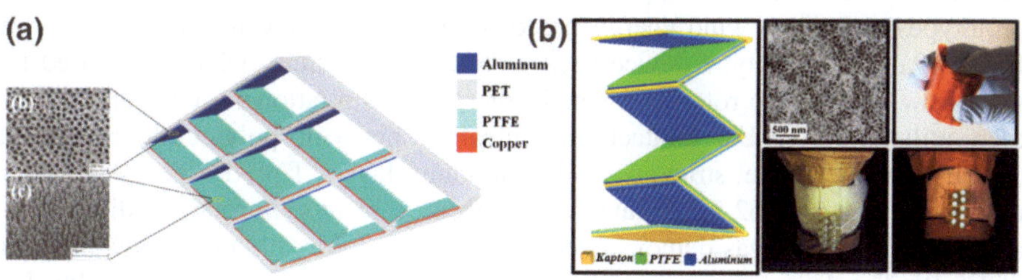

Fig. 2.17 **a** Rhombus shaped stacked structure to harvest mechanical energy from vibrations generated by human walking [38]. **b** Schematic diagram of the zigzag shaped multilayered triboelectric energy harvester; SEM micrograph of Al nanopores used one of the triboelectric layers; complete fabricated prototype of integrated multilayered triboelectric energy harvester [47]

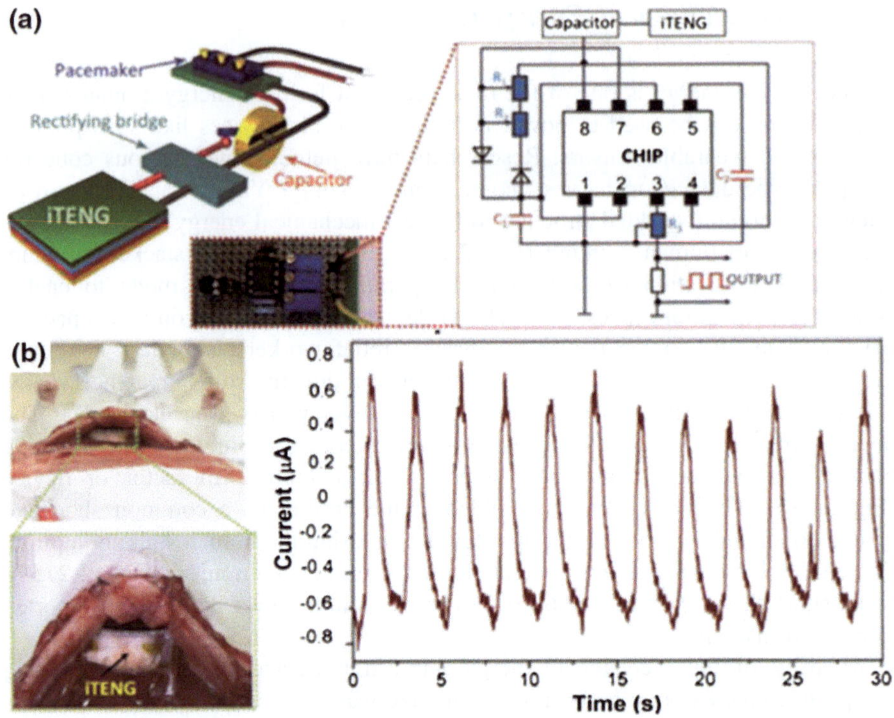

Fig. 2.18 Schematic diagram of the breathing driven TENG to power a pacemaker device [54]

2.4.2 *Wind Based Energy Harvesters*

Wind energy is a natural form of energy which has been used as a renewable source of energy to generate energy at a large scale. Conventional wind energy turbines used for large scale wind energy harvesting are bulky and require a huge capital investment to set up. Triboelectric mechanism can be used to generate electrical energy using wind energy at a smaller scale, which can be used to power wireless sensor nodes and electrical appliances. Xie et al. [55] demonstrated one of the first prototypes to harvest wind energy based on triboelectric mechanism as shown in Fig. 2.19a. The prototype device used a wind cup structure, which converted the wind flow energy into rotational kinetic energy. This rotational kinetic energy was then utilized to enable contact electrification between polytetrafluoroethylene (PTFE) nanowire-like structure and aluminum layer. The device generated a maximum power of 62.5 mW at a wind speed of 15 m/s. A rotary sliding mechanism based TENG was demonstrated by Zhu et al. [56] with a high power output of 1.5 W. The device comprised of a rotator and stator as shown in Fig. 2.19b. The stator constituted of a fluorinated ethylene propylene (FEP) layer with micro-patterned complementary sectors as electrode. The rotor was fabricated using

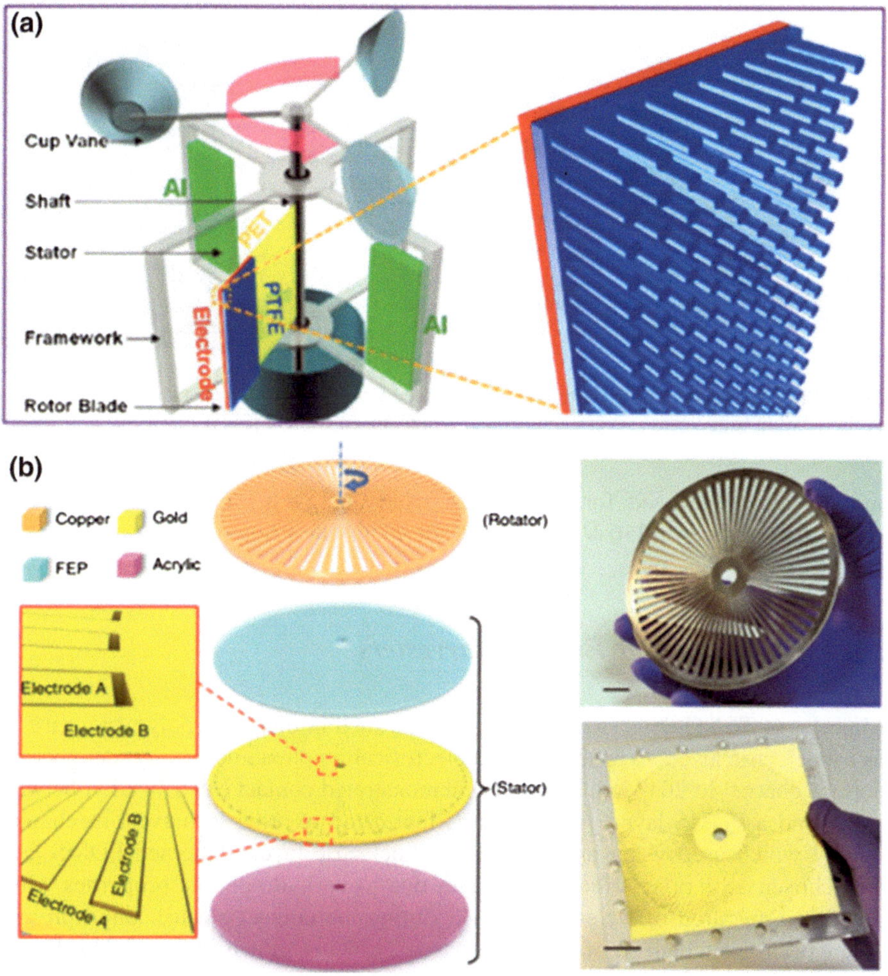

Fig. 2.19 **a** Wind-cup structure based TENG [55]. **b** High output TENG using micro-sector structures [56]

copper which acted as a triboelectric layer in conjunction with FEP. TENG was demonstrated to effectively harvest energy from air flow and water flow. Chen et al. [57] also demonstrated a rotary sliding mechanism based TENG which was shown to generate enough voltage for oxidation of sulphur dioxide, which has applications in self-powered air cleaning systems.

Another type of wind based TENG devices use flutter motion to convert wind energy into electrical energy. Bae et al. [58] proposed a TENG device which enabled contact electrification using fluttering motion of a flexible triboelectric layer. The flexible triboelectric layer was fabricated using a gold coated woven fabric as shown in Fig. 2.20a. The device demonstrated that the fluttering flag

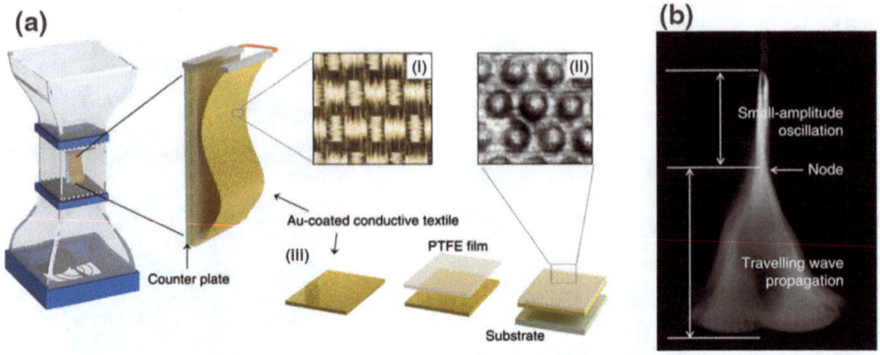

Fig. 2.20 **a** Schematic of the fluttering motion based wind energy harvester. **b** Fluttering motion of the flexible triboelectric layer captures using high speed camera [58]

motion can be utilized for small scale wind energy harvesting. It generated a maximum voltage of 250 V and current of 70 μA at a flow velocity of 22 ms^{-1}.

2.4.3 Water Based Energy Harvesters

Majority of TENG devices use solid materials such as dielectric polymers or metals as triboelectric layers to enable contact electrification phenomenon in order generate surface charge. In 2013, Lin et al. [59] demonstrated contact electrification between water and a PDMS layer. A device was fabricated in order to harvest mechanical energy which used water as one of the triboelectric layer. Patterned PDMS layer was chosen as another triboelectric layer due to its hydrophobic properties which aided in separation of water and PDMS during the contact electrification process. The working mechanism of the water based TENG is shown in Fig. 2.21a. During the contact electrification process due to contact separation motion between the two triboelectric layers enabled by external mechanical setup, water was observed to get positively charged as it donated electrons to PDMS layer. For the electrode corresponding to water as triboelectric layer, PMMA substrate coated with Cu was placed at the bottom of the experimental water tank. The experiments were conducted using deionized water, tap water and 0.6 M sodium chloride (NaCl) solution. The deionized water was shown to have a better performance as a triboelectric layer compared to tap water and NaCl solution. The performance for tap water and NaCl solution decreased due to the presence of electrolytes. As the water droplets cannot be completely prevented to form a layer on PDMS film, more positive charges including the dissolved ions in the water layer resulting into screening of negative surface charge generated on the PDMS film. Deionized water as a triboelectric layer along with patterned PDMS generated a charge density of 31.3 μQ m^{-2} during the contact electrification process. Lin et al. [60] fabricated a

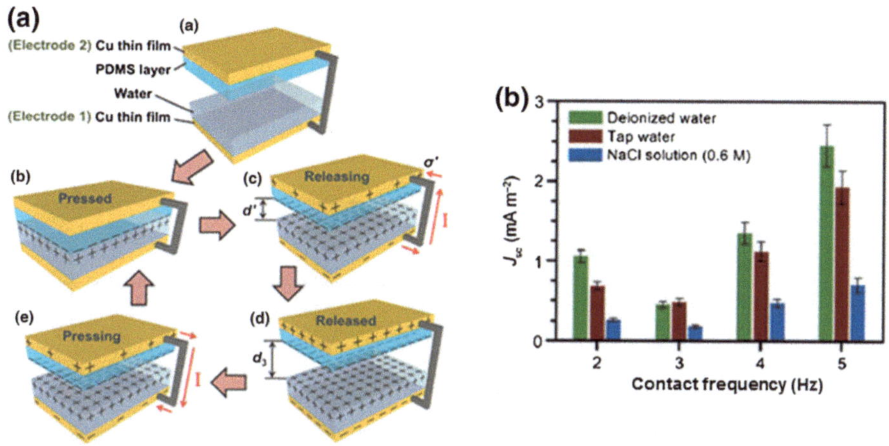

Fig. 2.21 **a** Operating mechanism of triboelectric energy harvester using water as triboelectric layer. **b** Current area density produced by different liquids as a triboelectric layer [59]

device to harvest energy using the charged water drops from rain or pipes. The device used PTFE hierarchal micro/nanostructures for contact electrification process with water drops. The device generated a peak voltage of 9.3 V and a peak current of 17 μA. Liang et al. [61] have demonstrated a prototype of transparent water based TENG which could be assembled on the windows in order to harvest energy from rainwater.

Apart from using water as a triboelectric layer, another method to harvest energy from water is utilizing the water's mechanical energy to enable contact electrification between two triboelectric layers. Yang et al. [62] demonstrated a TENG device, which used PTFE and polyamide (PA) as triboelectric layers. The device was fully enclosed in a spherical shell (Fig. 2.22) and was shown to harvest energy from water waves. Chen et al. [63] proposed the concept of network of small water wave energy harvester unit (Fig. 2.23a) to produce electricity at a large scale from sea/ocean waves. As per calculations, it was shown that power output of up to 1.15 MW can be generated from a water area of 1 km^2.

2.4.4 *Wearable Energy Harvesters*

Wearable energy harvesting devices which can harvest energy from human activities, are a solution to potentially power the wearable devices. One of the major direction of research in the area of wearable energy harvesters is focused towards textile or fabric based nanogenerators. Fabric based nanogenerators can camouflage with the clothes and also utilize large area of clothes for high power output. Kim et al. [64], demonstrated a stretchable weaved fabric for wearable energy harvesting applications. The fibers used to weave the fabric utilized PDMS tubes with high

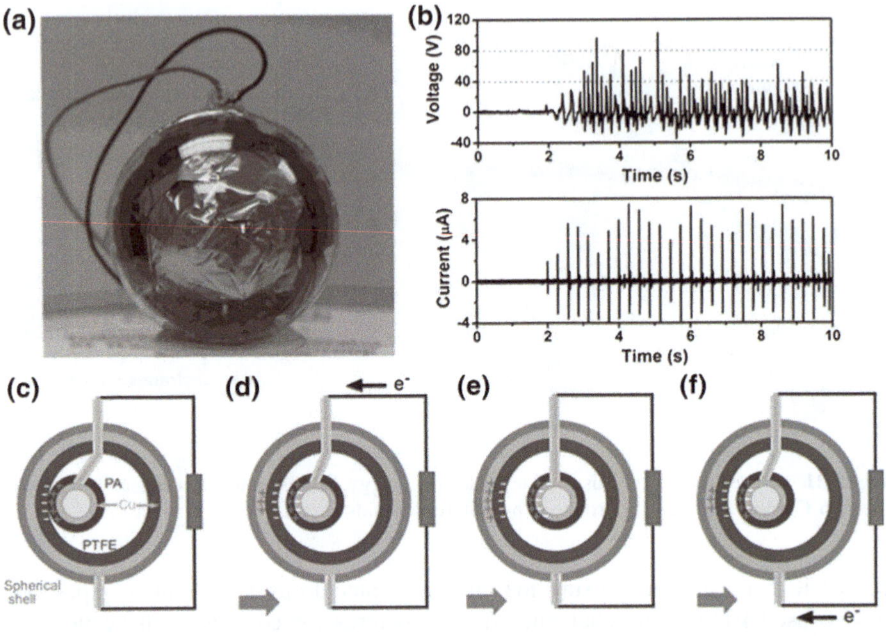

Fig. 2.22 **a** Prototype of fully enclosed TENG to harvest energy from water waves. **b** Voltage and current generated by the device when shaken by hand. **c–f** Operating mechanism of the TENG [62]

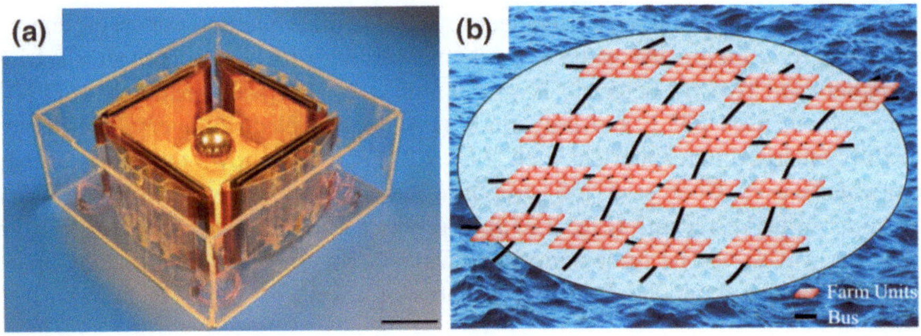

Fig. 2.23 **a** Single unit of water wave TENG. **b** A conceptual schematic of network or farm of unit TENGs to generate energy at a large scale [63]

aspect ratio nanostructured surface on the inner side and ZnO nanowires coated with Au as two triboelectric layers as shown in Fig. 2.24a. The weaved fabric with multiple fibers in an area of 14 cm × 14 cm generated a maximum voltage and current of 40 V and 210 μA, respectively. Seung et al. [65] demonstrated a fabric based TENG using a stacked structure of triboelectric layers as shown in Fig. 2.24b. A four layer stacked structure was shown to produce a peak voltage and current of 170 V and 120 μA, respectively.

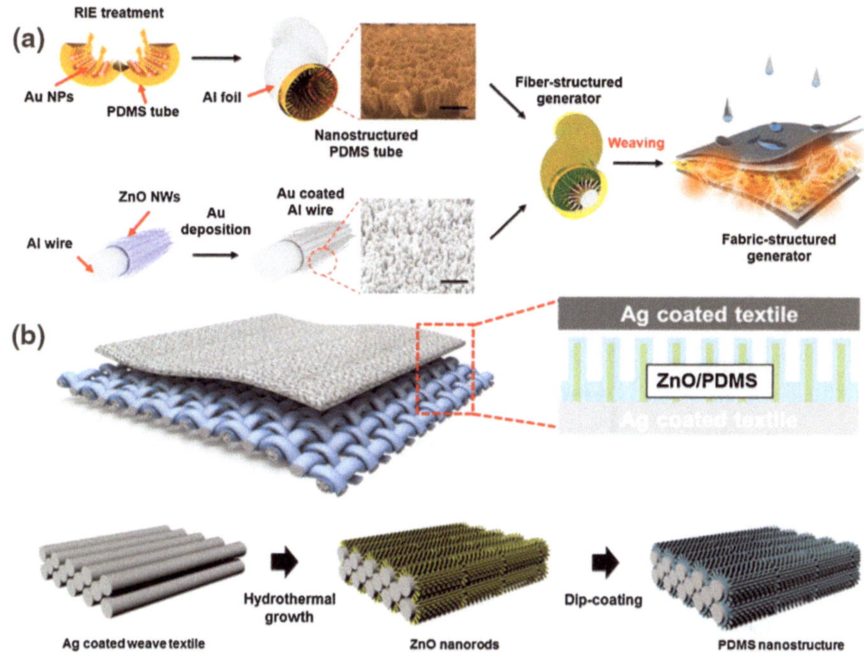

Fig. 2.24 **a** Fabric based TENG by weaving a fiber structured TENG [64]. **b** Fabric TENG by stacking of triboelectric layers [65]

Yang et al. [66] demonstrated a flexible TENG which utilized a wavy Kapton structure to enable contact electrification between a copper film and Kapton, as shown in Fig. 2.25. Due to the unique structure, the device could operate in both compressive and stretching mode. The device could generate a maximum voltage of up to 30 V from knee bending and straightening.

2.4.5 *Self-powered Sensors*

Sensors are an integral part of consumer and industrial electronic devices. Majority of these sensors work on the principle of measuring the change in properties such as capacitance and resistance. These sensors require external power supply in order to function. On the other hand, mechanical sensors to sense force, pressure or vibrations based on piezoelectric mechanisms do not require any external power for sensing as they have an ability to generate an electrical signal in response to a mechanical stimulation. Similarly, triboelectric mechanism can also generate an electrical signal when stimulated with an external mechanical force. These sensors

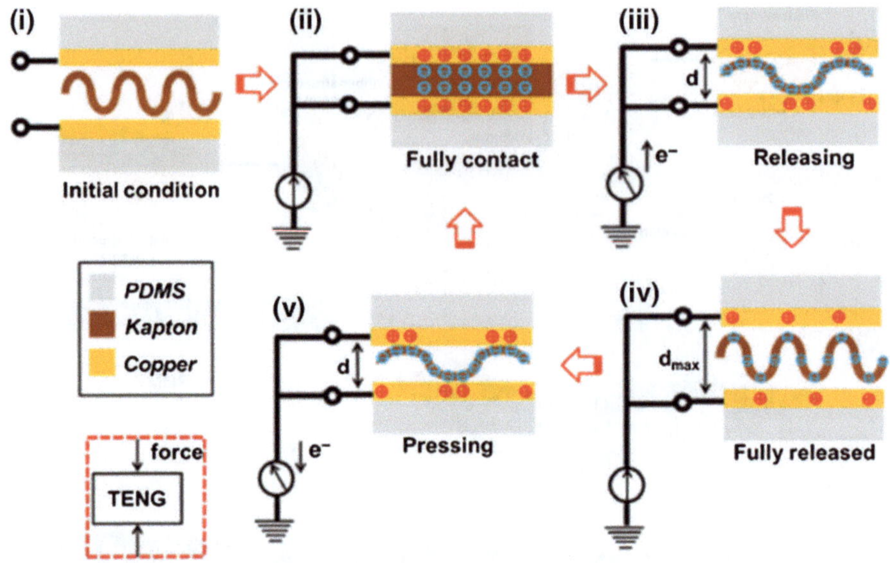

Fig. 2.25 Operating mechanism of the wavy structure based flexible TENG for wearable applications [66]

can be categorized as self-powered active sensors as they do not require an external power supply for sensing. Various kind of sensors have been realized by researchers using triboelectric mechanism and are reviewed in the following sections.

2.4.5.1 Tactile and Pressure Sensors

Yang et al. [67] demonstrated a self-powered tactile sensor array as shown in Fig. 2.26a. The device was designed for touch pad applications. It utilized micro-patterned PDMS as one of the triboelectric layers. As bare human finger touched the PDMS pixels, the electrons were injected from skin to PDMS due to contact electrification. Indium Tin Oxide (ITO) was used as an electrode beneath PDMS film to collect charged using triboelectric mechanism. The detection sensitivity of the pressure for every pixel was shown to be 0.29 ± 0.02 V kPa^{-1}. A similar device was also demonstrated by Meng et al. [68], which was shown to powered a liquid crystal display (LCD) screen.

Zhu et al. [69] fabricated a self-powered pressure sensor with a sensitivity of 44 V kPa^{-1} (Fig. 2.27a). The device was tested for durability and showed stable output for 10^5 cycles.

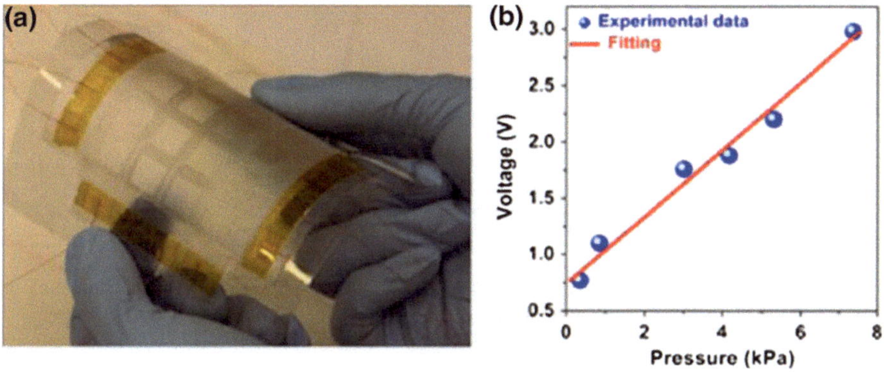

Fig. 2.26 **a** A photograph of the tactile sensor array. **b** Plot of the output voltage versus pressure [67]

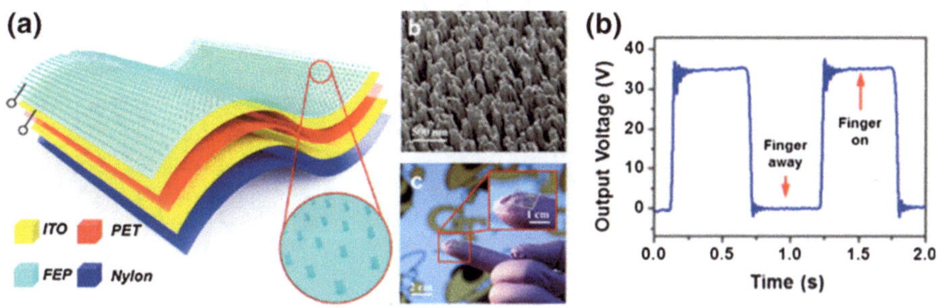

Fig. 2.27 **a** Schematic of the self-powered pressure sensor, SEM image of the polymer nanowires and a photograph of the fabricated sensor. **b** Output voltage produced by the sensor due to finger tapping [69]

2.4.5.2 Motion Tracking Sensors

Sensors based on triboelectric mechanism can also be utilized to sense motion parameters such as displacement, velocity and acceleration. Zhou et al. [70] proposed a grating based structure for motion sensing as shown in Fig. 2.28a. The structure comprised of parylene and silicon dioxide (SiO_2) gratings which underwent sliding motion relative to each other. Due to relative sliding motion of the two triboelectric layers, the two gratings experienced full overlap, partial separation and full separation. As the relative position of the two gratings changes, the open circuit potential between the ITO and Al electrode changes periodically. The data collected from the relative potential difference between electrodes and gap between the consecutive gratings was used to calculate the displacement with a resolution of 173 nm. Using the contact electrification effect and electrostatic induction between patterned electrodes, Wu et al. [71] designed a self-powered sensor for angular motion measurement (Fig. 2.28b).

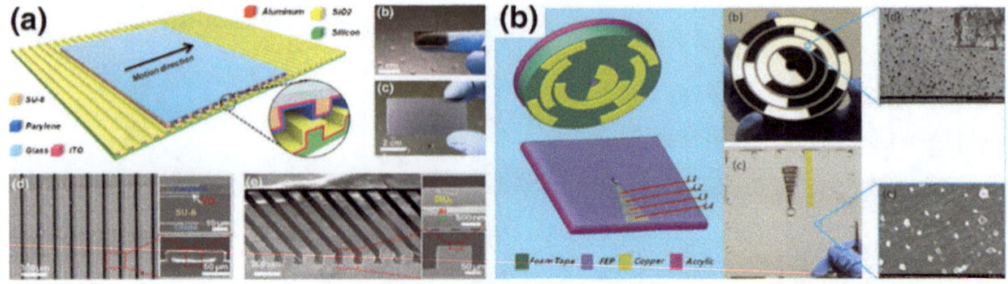

Fig. 2.28 **a** Triboelectric motion sensor using dual grating design [70]. **b** Schematic and photograph of the self-powered angular motion sensor [71]

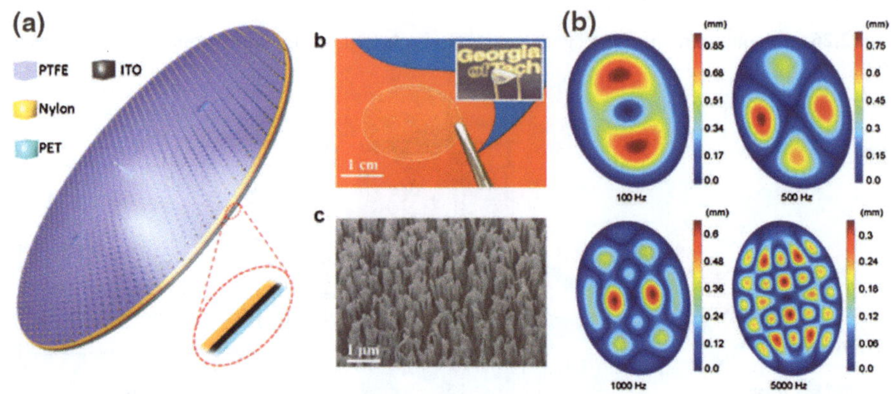

Fig. 2.29 **a** Eardrum inspired design of the voice recognition sensor. **b** Membrane vibrations simulated for different applied vibrations [72]

Various other self-powered mechanical sensors have been designed utilizing the triboelectric effect. Yang et al. [72] fabricated a self-powered voice recognition sensor using an eardrum inspired diaphragm design as shown in Fig. 2.29a. The sensor was also demonstrated for sensing arterial pulse wave for biometric recognition. Chen et al. [73] demonstrated a vibration sensor to sense vibration in the range of 0–200 Hz.

2.4.5.3 Chemical Sensors

Apart from mechanical sensing, triboelectric mechanism can also be used to detect chemical changes as the contact electrification phenomenon is affected by the chemical composition of the triboelectric layers. Several sensors have been demonstrated by the researchers using this principle. Lin et al. [74] detected the concentration of mercury ions (Hg^{2+}) with Tris-borate used as a buffer solution. The triboelectric layers comprised of a PDMS film and Au nanoparticles deposited onto an Au film. To enable the selective detection of Hg^{2+}, 3-mercaptopropionic acid

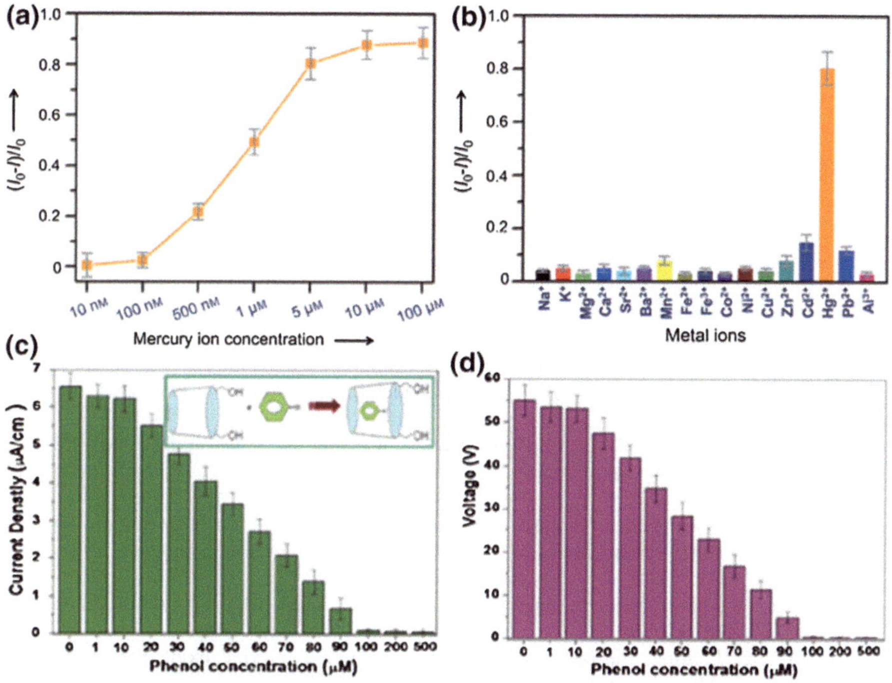

Fig. 2.30 **a** Sensitivity of the self-powered Hg^{2+} ion sensor. **b** Plot depicting the selectivity of the device for Hg^{2+}. The concentration of all metal ions tested was 5 μM [74]. **c** Current and **d** voltage signals from the sensor for various phenol concentrations [75]

(3-MPA) was self-assembled onto the surface of Au nanoparticles using strong Au-S interaction. The change in the performance of TENG resulted due to difference in triboelectric polarities of Au nanoparticles and Hg^{2+} ions. The device had a detection limit of 30 nM and linear range between 100 nM and 5 μM (Fig. 2.30a). The device was also shown to have high selectivity for Hg^{2+} ions. Li et al. [75] demonstrated a self-powered phenol detection sensor using titanium dioxide (TiO_2) nanowires and PTFE as triboelectric layers. β-cyclodextrin molecules were assembled onto TiO_2 nanowire layer, which acted as the phenol recognition element and also improved the performance of the contact electrification. The device achieved a detection sensitivity of 0.01 μM^{-1} by calibrating the current and voltage signals in the range of 10–100 μM.

2.5 Summary

In this chapter, various mechanical energy harvesting technologies have been reviewed. The applications of triboelectric devices in harvesting mechanical energy available in different forms has been reviewed. Thereafter, various self-powered devices based on triboelectric effect have been discussed including tactile sensing, motion sensing and chemical sensing.

References

1. T. Ikeda, *Fundamentals of Piezoelectricity* (Oxford University Press, Oxford, 1996)
2. S. Roundy, P.K. Wright, A piezoelectric vibration based generator for wireless electronics. Smart Mater. Struct. **13**, 1131 (2004)
3. N. White, P. Glynne-Jones, S. Beeby, A novel thick-film piezoelectric micro-generator. Smart Mater. Struct. **10**, 850–852 (2001)
4. M. Renaud, K. Karakaya, T. Sterken, P. Fiorini, C. Van Hoof, R. Puers, Fabrication, modelling and characterization of MEMS piezoelectric vibration harvesters. Sens. Actuators, A **145**, 380–386 (2008)
5. J.-Q. Liu, H.-B. Fang, Z.-Y. Xu, X.-H. Mao, X.-C. Shen, D. Chen et al., A MEMS-based piezoelectric power generator array for vibration energy harvesting. Microelectron. J. **39**, 802–806 (2008)
6. J. Kymissis, C. Kendall, J. Paradiso, N. Gershenfeld, Parasitic Power Harvesting in Shoes, in *Digest of Papers. Second International Symposium on Wearable Computers 1998*, 1998, pp. 132–139
7. S. Priya, Modeling of electric energy harvesting using piezoelectric windmill. Appl. Phys. Lett. **87**, 184101 (2005)
8. Z.L. Wang, J. Song, Piezoelectric nanogenerators based on zinc oxide nanowire arrays. Science **312**, 242–246 (2006)
9. X. Wang, J. Song, J. Liu, Z.L. Wang, Direct-current nanogenerator driven by ultrasonic waves. Science **316**, 102–105 (2007)
10. S.Y. Chung, S. Kim, J.H. Lee, K. Kim, S.W. Kim, C.Y. Kang et al., All-solution-processed flexible thin film piezoelectric nanogenerator. Adv. Mater. **24**, 6022–6027 (2012)
11. S. Lee, S.H. Bae, L. Lin, Y. Yang, C. Park, S.W. Kim et al., Super-flexible nanogenerator for energy harvesting from gentle wind and as an active deformation sensor. Adv. Funct. Mater. **23**, 2445–2449 (2013)
12. L. Lin, Y. Hu, C. Xu, Y. Zhang, R. Zhang, X. Wen et al., Transparent flexible nanogenerator as self-powered sensor for transportation monitoring. Nano Energy **2**, 75–81 (2013)
13. C. Williams, C. Shearwood, M. Harradine, P. Mellor, T. Birch, R. Yates, Development of an Electromagnetic Micro-Generator, in *IEE Proceedings on Circuits, Devices and Systems,* 2001, pp. 337–342
14. J.M. Lee, S.C. Yuen, W.J. Li, P.H. Leong, Development of an AA Size Energy Transducer with Micro Resonators, in *ISCAS'03 Proceedings of the 2003 International Symposium on Circuits and Systems, 2003*, vol. 4, 2003, pp. IV-876–IV-879
15. N.N. Ching, H. Wong, W.J. Li, P.H. Leong, Z. Wen, A laser-micromachined multi-modal resonating power transducer for wireless sensing systems. Sens. Actuators, A **97**, 685–690 (2002)
16. C. Cepnik, U. Wallrabe, A flat high performance micro energy harvester based on a serpentine coil with a single winding, in *Solid-State Sensors, Actuators and Microsystems Conference (TRANSDUCERS), 2011 16th International*, 2011, pp. 661–664
17. B. Yang, C. Lee, W. Xiang, J. Xie, J.H. He, R.K. Kotlanka et al., Electromagnetic energy harvesting from vibrations of multiple frequencies. J. Micromech. Microeng. **19**, 035001 (2009)
18. S. Kulkarni, E. Koukharenko, R. Torah, J. Tudor, S. Beeby, T. O'Donnell et al., Design, fabrication and test of integrated micro-scale vibration-based electromagnetic generator. Sens. Actuators, A **145**, 336–342 (2008)

19. J.M. Donelan, Q. Li, V. Naing, J. Hoffer, D. Weber, A.D. Kuo, Biomechanical energy harvesting: generating electricity during walking with minimal user effort. Science **319**, 807–810 (2008)
20. E. Halvorsen, S.D. Nguyen, MEMS Electrostatic Energy Harvesters with Nonlinear Springs, in *Advances in Energy Harvesting Methods*, (Springer, Berlin, 2013), pp. 63–90
21. S. Meninger, J.O. Mur-Miranda, R. Amirtharajah, A.P. Chandrakasan, J.H. Lang, Vibration-to-electric energy conversion. IEEE Trans. Very Large Scale Integ. (VLSI) Syst. **9**, 64–76 (2001)
22. D. Hoffmann, B. Folkmer, Y. Manoli, Analysis and characterization of triangular electrode structures for electrostatic energy harvesting. J. Micromech. Microeng. **21**, 104002 (2011)
23. B. Yang, C. Lee, R.K. Kotlanka, J. Xie, S.P. Lim, A MEMS rotary comb mechanism for harvesting the kinetic energy of planar vibrations. J. Micromech. Microeng. **20**, 065017 (2010)
24. D. Nguyen, E. Halvorsen, G. Jensen, A. Vogl, Fabrication and characterization of a wideband MEMS energy harvester utilizing nonlinear springs. J. Micromech. Microeng. **20**, 125009 (2010)
25. S.D. Nguyen, E. Halvorsen, Nonlinear springs for bandwidth-tolerant vibration energy harvesting. J. Microelectromech. Syst. **20**, 1225–1227 (2011)
26. J. Boland, Y.-H. Chao, Y. Suzuki, Y. Tai, Micro Electret Power Generator, in *IEEE the Sixteenth Annual International Conference on Micro Electro Mechanical Systems, 2003. MEMS-03 Kyoto*, 2003, pp. 538–541
27. T. Tsutsumino, Y. Suzuki, N. Kasagi, Y. Sakane, Seismic power generator using high-performance polymer electret, in *19th IEEE International Conference on Micro Electro Mechanical Systems, 2006. MEMS 2006 Istanbul*, 2006, pp. 98–101
28. Y. Sakane, Y. Suzuki, N. Kasagi, The development of a high-performance perfluorinated polymer electret and its application to micro power generation. J. Micromech. Microeng. **18**, 104011 (2008)
29. S. Matsusaka, H. Maruyama, T. Matsuyama, M. Ghadiri, Triboelectric charging of powders: a review. Chem. Eng. Sci. **65**, 5781–5807 (2010)
30. W. Harper, The Volta effect as a cause of static electrification, in *Proceedings of the Royal Society of London. Series A. Mathematical and Physical Sciences,* vol. 205, 1951, pp. 83–103
31. D. Davies, Charge generation on dielectric surfaces. J. Phys. D Appl. Phys. **2**, 1533 (1969)
32. Y. Murata, S. Kittaka, Evidence of electron transfer as the mechanism of static charge generation by contact of polymers with metals. Jpn. J. Appl. Phys. **18**, 421 (1979)
33. J. Henniker, Triboelectricity in polymers. Nature **196**, 474 (1962)
34. A. Diaz, R. Felix-Navarro, A semi-quantitative tribo-electric series for polymeric materials: the influence of chemical structure and properties. J. Electrostat. **62**, 277–290 (2004)
35. S. Niu, S. Wang, L. Lin, Y. Liu, Y.S. Zhou, Y. Hu et al., Theoretical study of contact-mode triboelectric nanogenerators as an effective power source. Energy Environ. Sci. **6**, 3576–3583 (2013)
36. S. Niu, Y. Liu, S. Wang, L. Lin, Y.S. Zhou, Y. Hu et al., Theory of sliding-mode triboelectric nanogenerators. Adv. Mater. **25**, 6184–6193 (2013)
37. Z.-H. Lin, Y. Xie, Y. Yang, S. Wang, G. Zhu, Z.L. Wang, Enhanced triboelectric nanogenerators and triboelectric nanosensor using chemically modified TiO_2 nanomaterials. ACS Nano **7**, 4554–4560 (2013)
38. W. Yang, J. Chen, G. Zhu, J. Yang, P. Bai, Y. Su et al., Harvesting energy from the natural vibration of human walking. ACS Nano **7**, 11317–11324 (2013)
39. G. Cheng, Z.-H. Lin, L. Lin, Z.-L. Du, Z.L. Wang, Pulsed nanogenerator with huge instantaneous output power density. ACS Nano **7**, 7383–7391 (2013)
40. M. Han, X.-S. Zhang, B. Meng, W. Liu, W. Tang, X. Sun et al., r-Shaped hybrid nanogenerator with enhanced piezoelectricity. ACS Nano **7**, 8554–8560 (2013)
41. Y. Yang, H. Zhang, Y. Liu, Z.-H. Lin, S. Lee, Z. Lin et al., Silicon-based hybrid energy cell for self-powered electrodegradation and personal electronics. ACS Nano **7**, 2808–2813 (2013)

42. Y. Yang, H. Zhang, J. Chen, Q. Jing, Y.S. Zhou, X. Wen et al., Single-electrode-based sliding triboelectric nanogenerator for self-powered displacement vector sensor system. ACS Nano **7**, 7342–7351 (2013)
43. L. Lin, Y. Xie, S. Wang, W. Wu, S. Niu, X. Wen et al., Triboelectric active sensor array for self-powered static and dynamic pressure detection and tactile imaging. ACS Nano **7**, 8266–8274 (2013)
44. Y. Yang, G. Zhu, H. Zhang, J. Chen, X. Zhong, Z.-H. Lin et al., Triboelectric nanogenerator for harvesting wind energy and as self-powered wind vector sensor system. ACS Nano **7**, 9461–9468 (2013)
45. Q. Jing, G. Zhu, P. Bai, Y. Xie, J. Chen, R.P. Han et al., Case-encapsulated triboelectric nanogenerator for harvesting energy from reciprocating sliding motion. ACS Nano **8**, 3836–3842 (2014)
46. Y. Yang, L. Lin, Y. Zhang, Q. Jing, T.-C. Hou, Z.L. Wang, Self-powered magnetic sensor based on a triboelectric nanogenerator. ACS Nano **6**, 10378–10383 (2012)
47. P. Bai, G. Zhu, Z.-H. Lin, Q. Jing, J. Chen, G. Zhang et al., Integrated multilayered triboelectric nanogenerator for harvesting biomechanical energy from human motions. ACS Nano **7**, 3713–3719 (2013)
48. W. Du, X. Han, L. Lin, M. Chen, X. Li, C. Pan et al., A three dimensional multi-layered sliding triboelectric nanogenerator, Adv. Energy Mater. 4(11), (2014)
49. H. Guo, X. He, J. Zhong, Q. Zhong, Q. Leng, C. Hu et al., A nanogenerator for harvesting airflow energy and light energy. J. Mater. Chem. A **2**, 2079–2087 (2014)
50. G. Zhu, C. Pan, W. Guo, C.-Y. Chen, Y. Zhou, R. Yu et al., Triboelectric-generator-driven pulse electrodeposition for micropatterning. Nano Lett. **12**, 4960–4965 (2012)
51. T.-C. Hou, Y. Yang, H. Zhang, J. Chen, L.-J. Chen, Z. Lin Wang, Triboelectric nanogenerator built inside shoe insole for harvesting walking energy. Nano Energy **2**, 856–862 (2013)
52. G. Zhu, Z.-H. Lin, Q. Jing, P. Bai, C. Pan, Y. Yang et al., Toward large-scale energy harvesting by a nanoparticle-enhanced triboelectric nanogenerator. Nano Lett. **13**, 847–853 (2013)
53. F.-R. Fan, L. Lin, G. Zhu, W. Wu, R. Zhang, Z.L. Wang, Transparent triboelectric nanogenerators and self-powered pressure sensors based on micropatterned plastic films. Nano Lett. **12**, 3109–3114 (2012)
54. Q. Zheng, B. Shi, F. Fan, X. Wang, L. Yan, W. Yuan et al., In vivo powering of pacemaker by breathing-driven implanted triboelectric nanogenerator. Adv. Mater. **26**, 5851–5856 (2014)
55. Y. Xie, S. Wang, L. Lin, Q. Jing, Z.-H. Lin, S. Niu et al., Rotary triboelectric nanogenerator based on a hybridized mechanism for harvesting wind energy. ACS Nano **7**, 7119–7125 (2013)
56. G. Zhu, J. Chen, T. Zhang, Q. Jing, Z.L. Wang, Radial-arrayed rotary electrification for high performance triboelectric generator. Nat. Commun. **5** (2014)
57. S. Chen, C. Gao, W. Tang, H. Zhu, Y. Han, Q. Jiang et al., Self-powered cleaning of air pollution by wind driven triboelectric nanogenerator, Nano Energy **14**, 217–225 (2014)
58. J. Bae, J. Lee, S. Kim, J. Ha, B.-S. Lee, Y. Park et al., Flutter-driven triboelectrification for harvesting wind energy. Nat. Commun. **5** (2014)
59. Z.H. Lin, G. Cheng, L. Lin, S. Lee, Z.L. Wang, Water–solid surface contact electrification and its use for harvesting liquid-wave energy. Angew. Chem. Int. Ed. **52**, 12545–12549 (2013)
60. Z.H. Lin, G. Cheng, S. Lee, K.C. Pradel, Z.L. Wang, Harvesting water drop energy by a sequential contact-electrification and electrostatic-induction process. Adv. Mater. **26**, 4690–4696 (2014)
61. Q. Liang, X. Yan, Y. Gu, K. Zhang, M. Liang, S. Lu et al., Highly transparent triboelectric nanogenerator for harvesting water-related energy reinforced by antireflection coating. Sci. Rep. **5** (2015)
62. Y. Yang, H. Zhang, R. Liu, X. Wen, T.C. Hou, Z.L. Wang, Fully enclosed triboelectric nanogenerators for applications in water and harsh environments. Adv. Energy Mater. **3**, 1563–1568 (2013)

63. J. Chen, J. Yang, Z. Li, X. Fan, Y. Zi, Q. Jing et al., Networks of triboelectric nanogenerators for harvesting water wave energy: a potential approach toward blue energy. ACS Nano **9**, 3324–3331 (2015)
64. K.N. Kim, J. Chun, J.W. Kim, K.Y. Lee, J.-U. Park, S.-W. Kim et al., Highly stretchable two-dimensional fabrics for wearable triboelectric nanogenerator under harsh environments. ACS Nano 2015
65. W. Seung, M.K. Gupta, K.Y. Lee, K.-S. Shin, J.-H. Lee, T.Y. Kim et al., Nanopatterned textile-based wearable triboelectric nanogenerator. ACS Nano **9**, 3501–3509 (2015)
66. P.K. Yang, L. Lin, F. Yi, X. Li, K.C. Pradel, Y. Zi et al., A flexible, stretchable and shape adaptive approach for versatile energy conversion and self powered biomedical monitoring. Adv. Mater. **27**, 3817–3824 (2015)
67. Y. Yang, H. Zhang, Z.-H. Lin, Y.S. Zhou, Q. Jing, Y. Su et al., Human skin based triboelectric nanogenerators for harvesting biomechanical energy and as self-powered active tactile sensor system. ACS Nano **7**, 9213–9222 (2013)
68. B. Meng, W. Tang, Z.-H. Too, X. Zhang, M. Han, W. Liu et al., A transparent single-friction-surface triboelectric generator and self-powered touch sensor. Energy Environ. Sci. **6**, 3235–3240 (2013)
69. G. Zhu, W.Q. Yang, T. Zhang, Q. Jing, J. Chen, Y.S. Zhou et al., Self-powered, ultrasensitive, flexible tactile sensors based on contact electrification. Nano Lett. **14**, 3208–3213 (2014)
70. Y.S. Zhou, G. Zhu, S. Niu, Y. Liu, P. Bai, Q. Jing et al., Nanometer resolution self-powered static and dynamic motion sensor based on micro-grated triboelectrification. Adv. Mater. **26**, 1719–1724 (2014)
71. Y. Wu, Q. Jing, J. Chen, P. Bai, J. Bai, G. Zhu et al., A self-powered angle measurement sensor based on triboelectric nanogenerator. Adv. Funct. Mater. **25**, 2166–2174 (2015)
72. J. Yang, J. Chen, Y. Su, Q. Jing, Z. Li, F. Yi et al., Eardrum-inspired active sensors for self-powered cardiovascular system characterization and throat-attached anti-interference voice recognition. Adv. Mater. **27**, 1316–1326 (2015)
73. J. Chen, G. Zhu, W. Yang, Q. Jing, P. Bai, Y. Yang et al., Harmonic-resonator-based triboelectric nanogenerator as a sustainable power source and a self-powered active vibration sensor. Adv. Mater. **25**, 6094–6099 (2013)
74. Z.H. Lin, G. Zhu, Y.S. Zhou, Y. Yang, P. Bai, J. Chen et al., A self-powered triboelectric nanosensor for mercury ion detection. Angew. Chem. Int. Ed. **52**, 5065–5069 (2013)
75. Z. Li, J. Chen, J. Yang, Y. Su, X. Fan, Y. Wu et al., β-cyclodextrin enhanced triboelectrification for self-powered phenol detection and electrochemical degradation. Energy Environ. Sci. **8**, 887–896 (2015)

Chapter 3
Study of Effect of Topography on Triboelectric Nanogenerator Performance Using Patterned Arrays

3.1 Motivation

Triboelectric effect has been recently demonstrated as a potential method to harvest mechanical energy from various sources in the surroundings which will otherwise be go wasted. Triboelectric effect is a type of contact electrification that results in the generation of surface charge as two different materials are put in contact with each other. The charging of surfaces occurs due to difference in tendencies of the materials to attract and donate electrons. Different tendencies of materials to attract or donate electrons arises due to chemical property of material known as electronegativity. The polarity of charge generated on the surface are decided by the material's position in triboelectric series [1]. During the device design stage, the materials for triboelectric layers should be chosen to have large difference in tendency to attract and donate electrons. The higher the difference in the tendency of the triboelectric layers, the higher is the surface charge generated during contact electrification.

Another important factor in determining the performance of triboelectric mechanism is the surface topography of triboelectric layers. Surface topography plays an important role as it decides the effective contact area between the triboelectric layers. Fan et al. [2] demonstrated that patterned films have a significantly higher performance than unpatterned films. Although patterning of triboelectric layers is known to improve the performance of triboelectric nanogenerators (TENG), it is still unclear which type of array pattern arrangement results in better device performance. In this chapter, we have used different array configurations fabricated using lithography and molding process to study the effect of surface topography on TENG performance.

L. Dhakar, *Triboelectric Devices for Power Generation and Self-Powered Sensing Applications*, Springer Theses,
DOI 10.1007/978-981-10-3815-0_3

3.2 Cantilever Based TENG—I

3.2.1 Device Design

To understand the effect of topography on the performance of TENG, a cantilever based TENG was designed as shown in Fig. 3.1. The cantilever based design was used to study the effect of surface topographies on TENG performance. Cantilever based design was chosen for the study, as motion and impact of triboelectric layers could be accurately modeled using established theories. We designed two cantilever based devices for the purpose of our study.

The cantilever based device comprised of two parts: (i) top part and (ii) bottom part. The device is based on periodic contact and separation motion between top and bottom parts. For the cantilever based TENG—I, the top part comprised of a gold thin film coated on a glass substrate with polydimethylsiloxane (PDMS) layer on top of it. The bottom part constituted of an array of SU-8 micropillars on silicon substrate with gold thin film coated on top of it. The top part was attached to an aluminum cantilever which was fixed using a clamping assembly at one end as shown in Fig. 3.1. The bottom part was kept fixed when the device is in operation. The periodic contact and separation motion between top and bottom parts was realized using the vibrating motion of cantilever when excited by vibrations provided by a mechanical shaker. In practical applications, these vibrations can be obtained from human motion [3], household equipment [4], machine vibrations etc. During the contact-separation motion, the bottom part also acted as an amplitude limiter for cantilever motion which induces non-linearity resulting in broadband behavior of the cantilever based resonant TENG—I [5–9].

3.2.2 Working Mechanism

The working mechanism for energy harvesting cycle is described in Fig. 3.2. Initially, before providing the mechanical excitation, the top and bottom

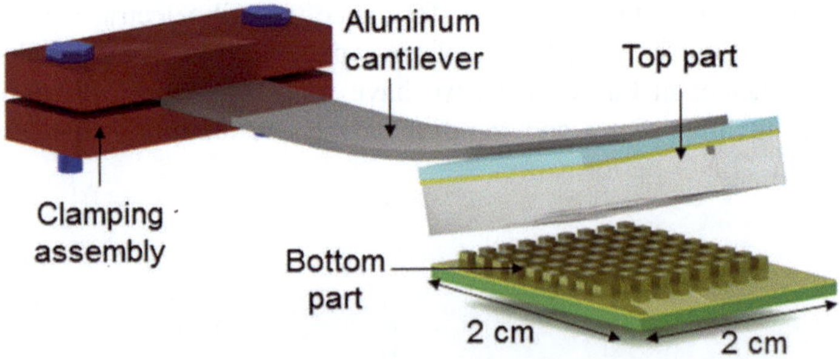

Fig. 3.1 Schematic of the cantilever based TENG—I

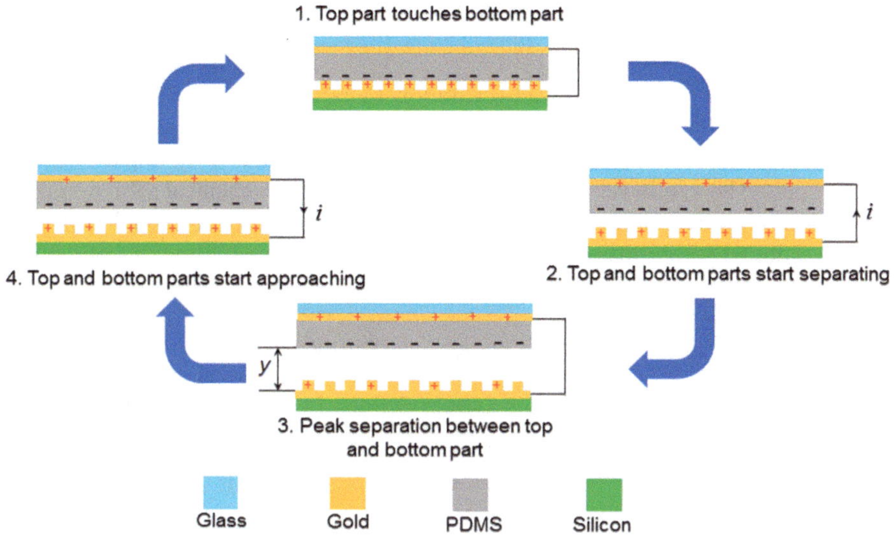

Fig. 3.2 Working mechanism of TENG at different stages of cantilever vibration

triboelectric layers are in separated position without any prior charges. As the device is mechanically excited with vibrational frequency in operating range, the cantilever starts to vibrate. Thereafter, the distance y as shown in Fig. 3.2 decreases, and top (PDMS) and bottom (gold) triboelectric layers come in contact with each other. As per triboelectric series, gold has a higher tendency to donate electrons as compared to PDMS. Therefore in state 1, when the gold coated on the SU-8 micropillars and PDMS layer are in contact, gold layer gets positively charged whereas the PDMS layer gets negatively charged. Now as the two parts start separating from each other, bottom gold electrode is at higher electric potential relative to the top gold electrode. Therefore, electrons start flowing from top gold electrode to bottom gold electrode resulting in current i as shown in state 2. The current keeps flowing till the cantilever reaches the highest displacement on the opposite side, where the separation between top and bottom triboelectric layers is maximum (state 3). At this point, an electrostatic equilibrium is reached. As the cantilever is vibrating under mechanical excitation, the top triboelectric layer starts approaching (state 4) the bottom layer and electrostatic equilibrium between the different layers is disturbed. Now, the current starts flowing from top electrode to bottom electrode, as opposed to the current direction during separation motion of two parts. Thereafter, electrostatic equilibrium is reached, when the gold layer and PDMS layer are again in contact with each other (state 1). The charges flowing in the external circuit connected to the TENG can be harvested to power wireless sensor nodes or low power electronic devices.

3.2.3 Theory

The cantilever is used for contact-separation motion to generate surface charges using contact electrification through triboelectric effect. This section discusses the mechanics of top part attached to the vibrating cantilever and its impact with the bottom part. The cantilever is excited with a sinusoidal mechanical vibration using an electromagnetic shaker. Cantilever vibrating motion can be modeled using a forced, damped spring mass system. The equation for forced vibrations of damped spring mass system can be written as:

$$m\ddot{y} + c\dot{y} + ky = F_0 \sin(\omega t) \tag{3.1}$$

The steady state solution for the system defined by Eq. (3.1) can be given by:

$$y = \frac{F_0}{\sqrt{(k - m\omega^2)^2 + (c\omega)^2}} \sin(\omega t - \phi) \tag{3.2}$$

where $\phi = \tan^{-1}(\frac{c\omega}{k - m\omega^2})$, m is mass, C is damping coefficient, k is the spring constant, F_0 is the amplitude of the sinusoidal excitation force, ω is the excitation frequency. Differentiating Eq. (3.2) with respect to time, the velocity function of the mass can be obtained as:

$$\dot{y} = \frac{F_0 \omega}{\sqrt{(k - m\omega^2)^2 + (c\omega)^2}} \cos(\omega t - \phi) \tag{3.3}$$

When the excitation frequency of the shaker matches the resonant frequency (ω_r) of the shaker, from Eq. (3.3) the equation of motion can be written as:

$$\dot{y} = \frac{F_0}{c} \cos(\omega_r t - \phi) \tag{3.4}$$

Figure 3.3 shows the vibrating cantilever beam impacting the bottom part. The average impact force (F) is generated due to the change in momentum of the cantilever which can be expressed as:

$$F = \frac{change\ in\ momentum}{\Delta T} = \frac{(m.v_n - 0)}{\Delta T} = \frac{m\dot{y}}{\Delta \mathrm{T}} \tag{3.5}$$

where, v_n is the normal component of velocity for the top part and ΔT is the duration of impact. In Eq. (3.5), for the force calculation it is assumed that the top part comes to rest after impacting the bottom part. The cantilever can be modeled as a spring with an attached proof mass. The spring-mass system undergoes forced oscillation due to the mechanical excitation provided by the shaker. As shown in the Fig. 3.4, the potential energy of the system is maximum at the two extreme

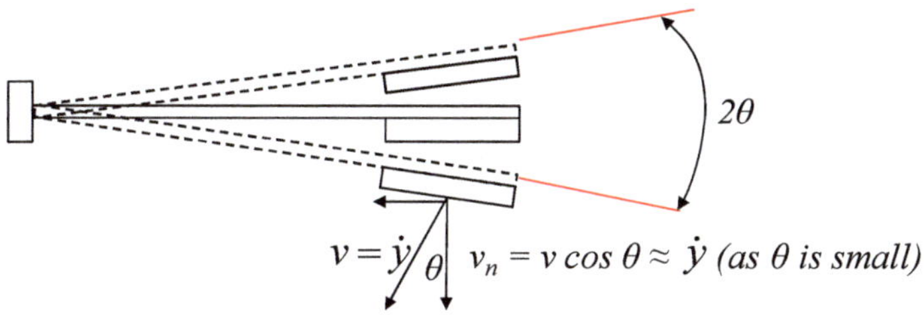

Fig. 3.3 Calculation of impact force between the *top* and *bottom* part

positions of the oscillation. The kinetic energy is maximum at the mean position when the spring does not have any potential energy stored in it. For the motion of the system, where the motion of proof mass is limited by a mechanical stopper, the potential energy of the system does not change as it is stored in form of stress energy in the cantilever. It is assumed that all the kinetic energy of the proof mass is converted to surface charges generated due to contact electrification, sound energy and heat energy.

After plugging the value of $\dot{y}$, Eq. (3.5), can be written as:

$$F = \frac{mF_0}{c\Delta T}\cos(\omega_r t - \phi) \tag{3.6}$$

This model predicts the impact force between top and bottom parts. It models the vibrating cantilever as a damped spring mass system and assumes that the top part comes to rest after impacting the bottom part. The local deformation for PDMS layer and individual SU-8 micropillar can be modeled using the Hertz Theory [10]. Keeping all the geometrical parameters and material properties same, the relation between duration of contact and impact velocity is given by [10, 11]:

$$\Delta T \propto (\dot{y})^{-1/5} \tag{3.7}$$

Fig. 3.4 Potential energy (P.E.) and kinetic energy (K.E.) of the oscillating system at different positions

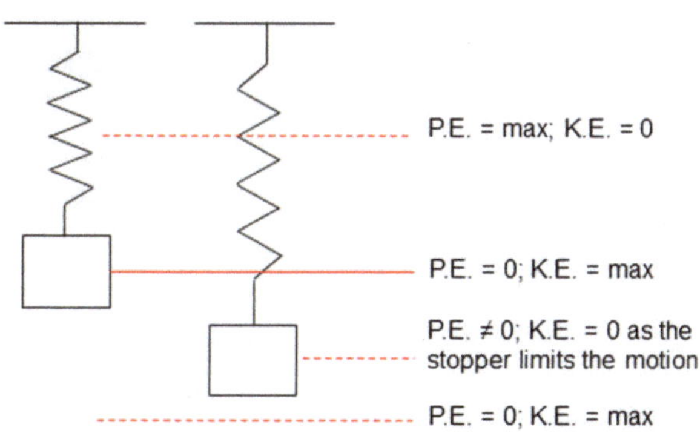

where, $\dot{y}$ is the impact velocity. From Eqs. (3.6) and (3.7), following relation can be observed:

$$F \propto F_0^{6/5} \tag{3.8}$$

It is clear from Eq. (3.8), that as the magnitude of the excitation force or input acceleration is increased, the impact force between top and bottom parts increases. This simplified model of impact motion serves the purpose of understanding the qualitative relation between the magnitude of excitation force and impact force between top and bottom parts.

The micropillar array can be assumed to be an array of springs fixed at one end and free at the other end (Fig. 3.5a). As Euler theory cannot be applied because the slender shape assumption is not valid for the pillar shape [12], Hooke's law is used to calculate the spring constant of the SU-8 micropillars. Therefore, the spring constant using Hooke's law is given by following equations:

$$\sigma = E\varepsilon \tag{3.9}$$

$$\frac{F}{A} = E\frac{\Delta l}{l} \tag{3.10}$$

$$k = \frac{F}{\Delta l} = \frac{EA}{l} \tag{3.11}$$

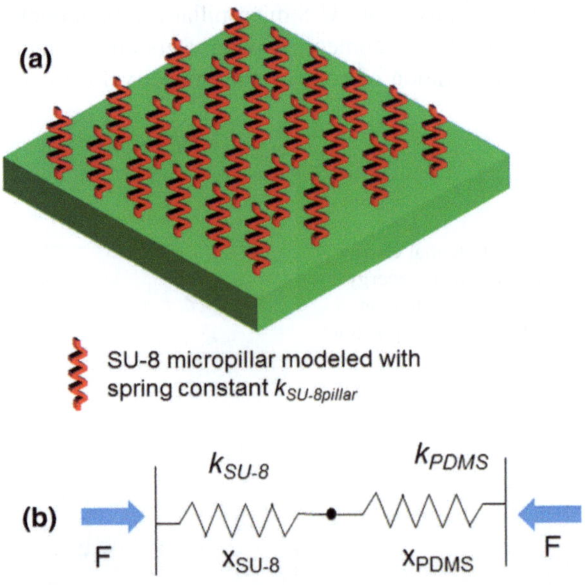

Fig. 3.5 **a** SU-8 micropillars modeled as linear springs. **b** In contact state, deformation PDMS layer and SU-8 micropillars is calculated by spring in series model

Table 3.1 Parameters for spring constant calculation

Structure	Young's modulus	Dimension
SU-8 pillar	~2 GPa	50 μm × 50 μm × 50 μm
PDMS layer	~500 kPa	2 cm × 2 cm × 500 μm

where σ is stress, E is Young's modulus, F is the force applied in normal direction, A is the cross section area, l is the height of micropillar, Δl is the change in height and k is the spring constant.

The spring constant for the individual micropillar can be calculated using Eq. (3.11). For sample calculation, a micropillar dimension of 50 μm × 50 μm × 50 μm is used as summarized in Table 3.1. The spring constant for individual micropillar is calculated to be 100,000 N/m. All the SU-8 micropillars are connected in parallel during deformation in contact mode. Therefore the effective spring constant for SU-8 can be written as:

$$k_{SU-8} = n \times k_{SU-8pillar} \tag{3.12}$$

where n is the total number of pillars in the array.

Using Eq. (3.12) the value of k_{SU-8} is calculated to be 4×10^9 N/m as n = 40,000 for the micropillar array of 50 μm × 50 μm pillars. The spring constant for the PDMS layer can also be obtained using Eq. (3.11) and was calculated to be 4×10^5 N/m.

When SU-8 micropillar array and the PDMS layer are in contact, the deformation in both PDMS and SU-8 can be calculated using simple spring in series model as shown in Fig. 3.3b. For two springs in series as shown in Fig. 3.5b, the ratio of deformation in two springs can be calculated as:

$$\frac{x_{PDMS}}{x_{SU-8}} = \frac{k_{SU-8}}{k_{PDMS}} = \frac{4 \times 10^9}{4 \times 10^5} = 10000 \tag{3.13}$$

As can be seen from Eq. (3.13), during the contact state, most of the deformation occurs in PDMS layer. This is due to high stiffness of SU-8 pillars as compared to the PDMS layer. This indicates that the contact area between SU-8 micropillars and PDMS layers is due to two factors: (i) primary contact area because of micropillar cross section and (ii) contact area resulting due to local elastic deformation of PDMS layer. Deformation of SU-8 micropillars can be neglected and does not contribute significantly to the contact area between two materials. Figure 3.6 shows the results using finite element method which was performed using COMSOL. The figure depicts the deformation in the PDMS layer. It can be observed that apart from top contact area between PDMS and micropillar, the contact area also increases at the micropillar edges due to local deformation.

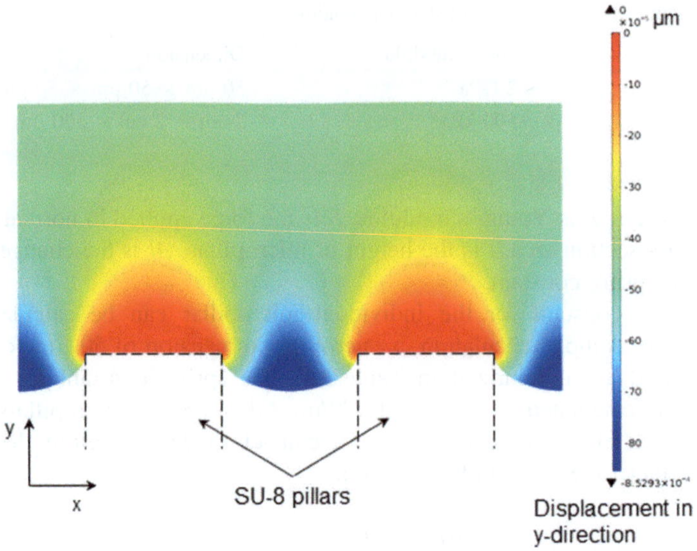

Fig. 3.6 Deformation observed in the PDMS layer using finite element method simulation

3.2.4 *Fabrication Process*

The steps involved in the fabrication of top and bottom parts are shown in Fig. 3.7. For the fabrication of top part, glass slide was cleaned in ultrasonic bath using acetone and isopropyl alcohol (IPA). A 100 nm thick film of gold is then coated on

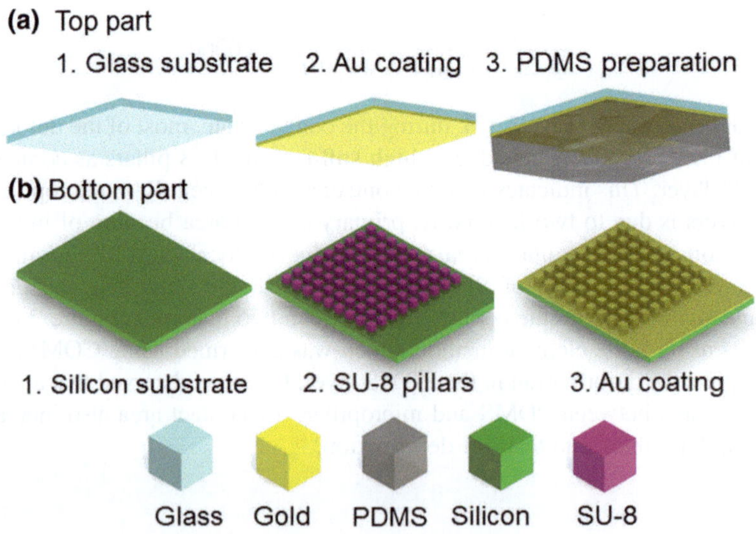

Fig. 3.7 Fabrication of the *top* and *bottom* part

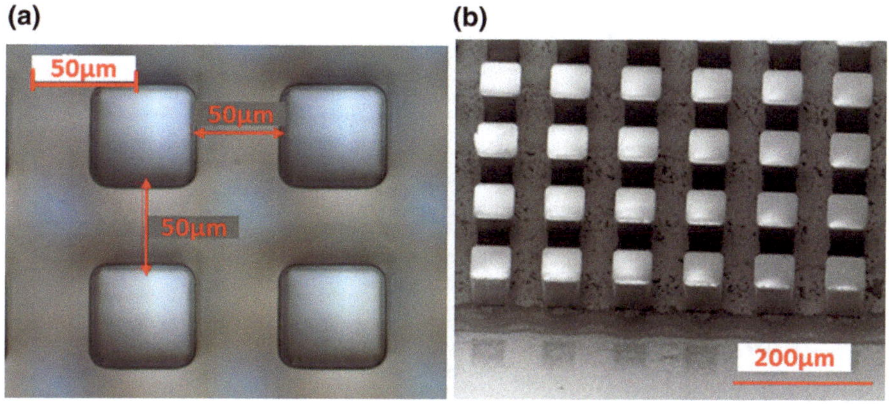

Fig. 3.8 **a** Optical and **b** SEM images of array of SU-8 micropillar array of dimension of 50 μm × 50 μm × 50 μm

the glass sample using thermal evaporation. This gold layer serves as the top electrode for TENG. On top of gold film, a 500 μm thick layer of 10:1 PDMS mixture was spin coated using SYLGARD® 184 silicone elastomer kit. The spin coated PDMS layer was then kept in oven at 80 °C for 2 h for curing. The PDMS layer acts as a dielectric layer through which the gold electrode gets charged by electrostatic induction mechanism. The top part was then attached to an aluminum cantilever as shown in Fig. 3.1 using epoxy adhesive.

The bottom part fabrication started with cleaning of silicon substrate. After cleaning, a 50 μm thick layer of negative photoresist SU-8 2025 from MicroChem was spin coated on the silicon substrate followed by soft bake in steps at 65 °C and 95 °C for 3 and 7 min, respectively. SU-8 was then exposed for patterning using photolithography followed by post exposure bake. The samples were then developed using MicroChem's SU-8 developer. The images of fabricated SU-8 micropillar arrays are shown in Fig. 3.8a, b. Thereafter, the micropillar arrays were hard baked. The SU-8 micropillars were then coated with a 100 nm gold film which served as the bottom electrode for cantilever TENG—I.

3.2.5 *Broadening of Operating Bandwidth Using Mechanical Stopper*

As the bottom part of cantilever TENG—I device acted as a mechanical stopper, the device demonstrated broadband characteristics due to spring stiffening, which is discussed in this section in detail. The spring constant variation of the composite cantilever due to stopper is shown in Fig. 3.9a. As the amplitude of the proof mass increases more than d, the bottom part obstructs its motion. Here d is the initial gap between the top and bottom triboelectric layer. Due to stopper, spring stiffening

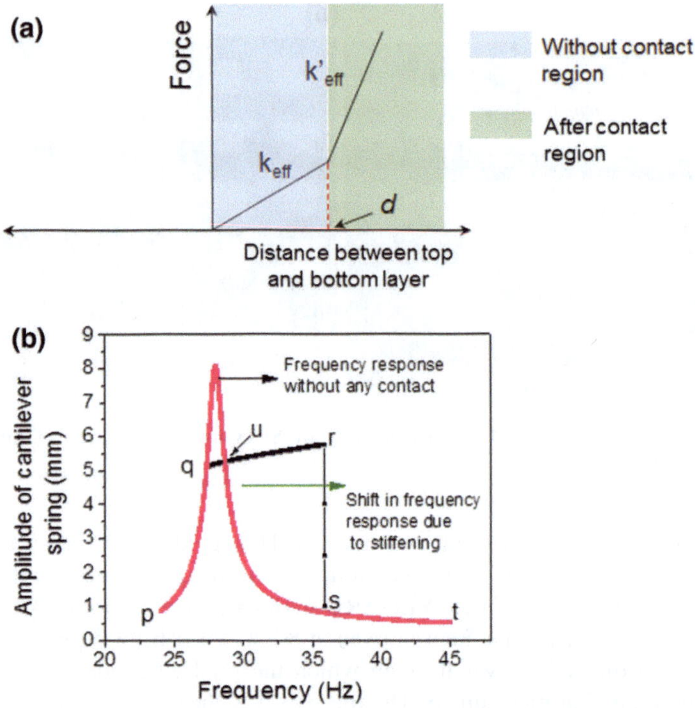

Fig. 3.9 **a** Variation of effective spring constant with respect to the displacement of proof mass as per piecewise linear model. **b** Simulated result for the amplitude of vibrating cantilever with an amplitude limiter

occurs and the effective spring constant increases from k_{eff} to k'_{eff}. In this case, the vibrating cantilever system can be considered as a piecewise linear oscillator, the equation of motion with dimensionless parameters is given by [13]:

$$\ddot{u} + 2\zeta_0 \dot{u} + u = r^2 \sin(r\tau) + f_1(u, \dot{u}) \tag{3.14}$$

where $\omega_0 = 2\pi f_0$; $\omega_1 = 2\pi f_1$; $\tau = \omega_0 t$; $r = \omega/\omega_0$; $r_1 = \omega_1/\omega_0$; $u = z/Y$; $\delta_1 = d/Y$; f_0 and f_1 are the frequency characteristics of cantilever TENG without and with stopper (bottom part) respectively; z is the displacement at the free end of the cantilever beam; Y is the amplitude of the mechanical excitation provided in the form of $y = Y \sin \omega t$; ζ_0 is the damping ratio for cantilever TENG. The expression for $f_1(u, \dot{u})$ is given by the following equation:

$$f_1(u, \dot{u}) = \begin{cases} -2r_1\zeta_1\dot{u} - r_1^2 u + r_1^2 \delta_1, & u \geq d \\ 0, & u < d \end{cases} \tag{3.15}$$

where ζ_1 is the damping ratio of stopper; and d is the distance of the top part from bottom part. The amplitude A of proof mass can be expressed as a function of dimensionless excitation frequency parameter r as:

$$\pi^2 r^4 = Z_1^2 + Z_2^2 \tag{3.16}$$

Z_1 and Z_2 are given by the following equations:

$$Z_1 = -2\zeta_0 Ar\pi - r_1\zeta_1 Ar(\pi - 2\varphi_1 - \sin 2\varphi_1) \tag{3.17}$$

$$Z_2 = \pi A\left(1 - r^2\right) - \left\{0.5r_1^2 A(2\varphi_1 - \sin 2\varphi_1 - \pi) + 2r_1^2\delta_1 cos\varphi_1\right\} \tag{3.18}$$

The frequency response given by Eqs. (3.14)–(3.17) is simulated using MATLAB for d = 5 mm at an acceleration of 1 g. In the simulation, the value for ζ_0, ζ_1, f_0 and f_1 are assumed to be 0.02, 0.12, 28 and 150 Hz, respectively. The simulation result is shown in Fig. 3.6b. The top part initially moves freely and follows the frequency response of a linear oscillator without stopper from point p to q. Once the top part moves more than d, it engages with the stopper (bottom part) and effective stiffness and damping coefficient increases. As the frequency of mechanical excitation force increases, motion of free end follows the piecewise linear model from point q to r. After reaching point r the amplitude abruptly decreases as the vibration amplitude drops to match with the response of linear system. The amplitude returns back to point s and follows the frequency response of that without any stopper from point s to point t. During the frequency down-sweep, the amplitude increases from point t to u, at which the cantilever starts engaging with the stopper. A sudden change in the amplitude occurs at point u and thereafter the response is identical to the response during up-sweep till the frequency reaches point q. From point q onwards, the cantilever disengages from the stopper and follows the frequency response same as the linear system.

3.2.6 *Output Voltage and Power*

A better understanding of the working mechanism of the device can be gained using simulation of potential distribution across the electrodes. The simulations were carried out using COMSOL®. In the simulations conducted, a charge density of $-10\ \text{nC/m}^2$ was assumed on the PDMS surface. The results are shown in Fig. 3.10 for different values of y which is the distance between the top and bottom part. The bottom electrode is taken as reference in the simulation and was grounded for the purpose. As can be seen from the simulation results, the potential difference between the top and bottom electrodes keeps increasing as the two electrodes are separated further away from each other. The potential difference for a distance of 1 mm is 0.8 V while it increases to more than 5.8 V as the distance is increased to

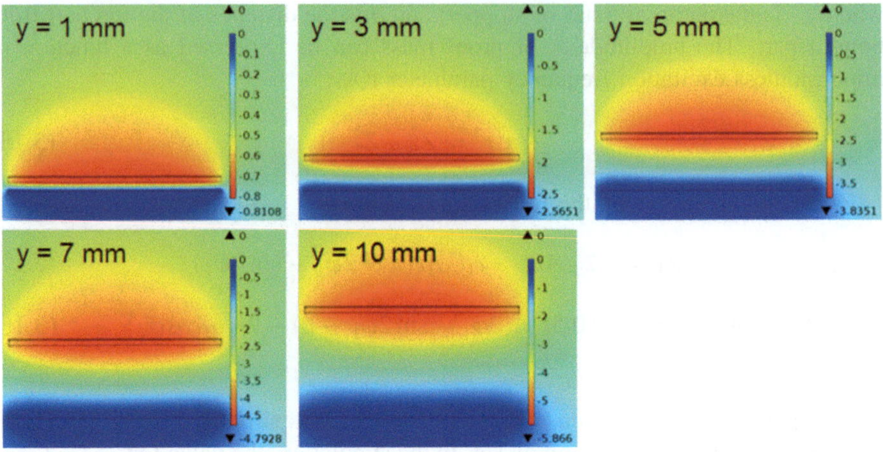

Fig. 3.10 Potential distribution at different distance between *top* and *bottom* parts along the normal direction. '*y*' is the gap between *top* and *bottom* triboelectric layers

10 mm. So as the top and bottom parts are further separated, mechanical work is done against the electric field due to which the potential difference is increased between the top and bottom electrodes. This results in the flow of electrons from the top electrode to the bottom electrode as the top and bottom triboelectric layers are separated.

3.2.7 Design of Experiment

To study the effect of micropillar array configuration, six sample variations for the SU-8 micropillar array were prepared. The overall sample size for all six micropillar arrays was 2 cm × 2 cm. The micropillar array configuration is shown in Fig. 3.11a. In the micropillar arrays, the gap between the micropillars is denoted as g. The cross section of an individual micropillar is characterized by the length l and width w. The height of the micropillars h and the gap g between individual SU-8 micropillars is kept constant for all the six micropillar arrays fabricated. Different micropillar sizes fabricated for the micropillar arrays are shown in Fig. 3.11b. An important parameter to characterize the micropillar arrays is fill factor, which is defined here as:

$$Fill\,factor = l/g \tag{3.19}$$

The fill factor defined above characterizes the ratio of length of micropillar and gap between the micropillars along the same direction keeping other parameters

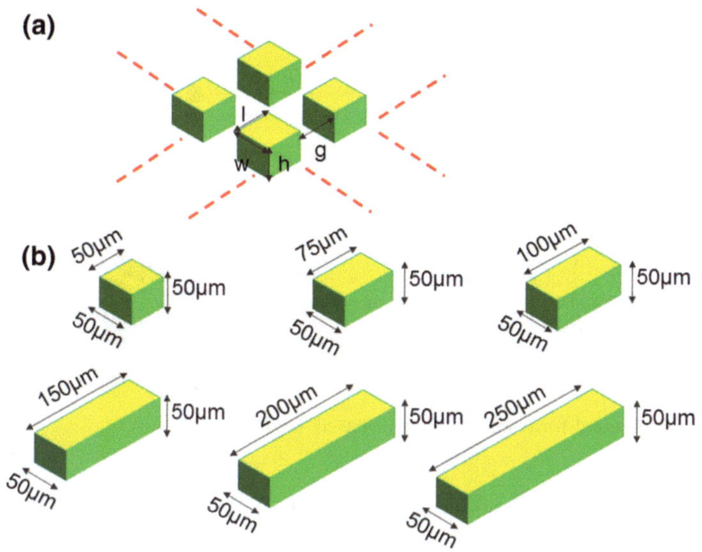

Fig. 3.11 **a** Micropillar array dimensional parameters **b** different pillar dimensions used for preparing SU-8 micropillar arrays

Table 3.2 Fill factors for micropillar arrays

Sample number	Pillar size	Fill factor
S1	50 μm × 50 μm	1
S2	75 μm × 50 μm	1.5
S3	100 μm × 50 μm	2
S4	150 μm × 50 μm	3
S5	200 μm × 50 μm	4
S6	250 μm × 50 μm	5

intact. The fill factor calculated for the six pillar arrays are summarized in Table 3.2.

3.2.8 Experimental Setup

To study the cantilever TENG devices, an electromagnetic shaker was used to provide mechanical excitation. The setup for testing the devices is shown in Fig. 3.12. A software was used to generate sinusoidal signal which was then sent to an amplifier to amplify the signal. The amplified signal was then sent to the electromagnetic shaker to provide vibrations. To measure the acceleration provided by the electromagnetic shaker, an accelerometer was assembled on shaker along with the device. The accelerometer signal was then again sent to the software resulting in

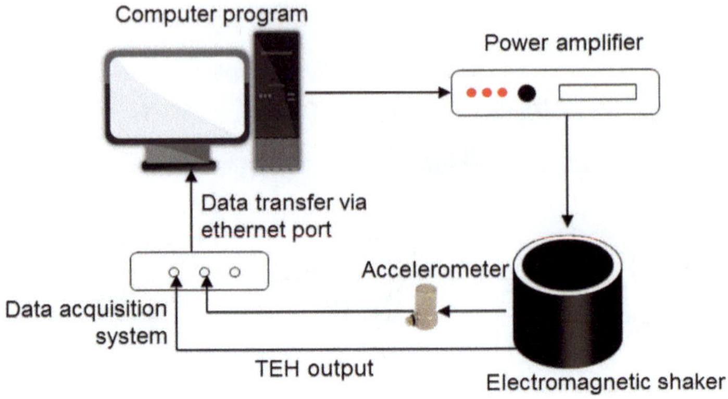

Fig. 3.12 Experimental for testing cantilever based TENG

a closed loop. This feedback loop was used to maintain the acceleration level provided by the shaker.

3.2.9 *Results and Discussion*

3.2.9.1 Effect of Increasing Acceleration

The effect of excitation acceleration was studied on the output open circuit voltage generated by cantilever TENG—I by varying the acceleration provided by shaker from 0.4 to 1.8 g in steps of 0.2 g. As the amplitude of excitation acceleration or force increases, the impact force also keeps increasing as discussed in earlier section and explained by Eq. (3.8). Higher impact force results in increased elastic deformation in the PDMS layer which leads to increased contact area between the two triboelectric layers. The increased contact area results in increased triboelectric generation and performance of the device. Figure 3.13 shows the time domain

Fig. 3.13 Output voltage of cantilever TENG—I at different acceleration levels from shaker

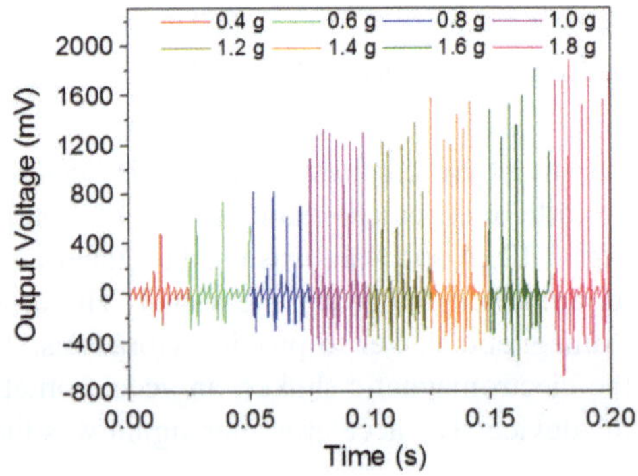

signal for the output voltage at different accelerations. The peak output voltage increases from 450 to 1900 mV as the acceleration is increased from 0.4 to 1.8 g.

3.2.9.2 Calculation of Charge Density

During the contact electrification process, triboelectric charges are generated due to which surfaces gets charged. These surface charges result in electrostatically induced charges on electrodes. The charge densities on electrodes can be calculated using the calculation of charge by obtaining the area under the curve for current versus time graph as given by:

$$q = \int_{t_a}^{t_b} i(t)dt \tag{3.20}$$

Figure 3.14a shows the current for cantilever TENG—I in operation. For the calculation of charge flowing through the external circuit, a peak was selected as highlighted in Fig. 3.14a. The magnified graph for the peak is shown in Fig. 3.14b. For calculating the charge flown through the external circuit as given by Eq. (3.20), current is integrated between time t_a and t_b. Area under the curve for current versus time graph between time t_a and t_b is calculated to obtain the charge flown. The charge flown through the external circuit between time t_a and t_b was calculated to be 1.48×10^{-10} °C. The charge density on the electrodes can be calculated to be 37 pC/cm^2 using the following equation:

$$\sigma = \frac{q}{Total\, electrode\, area} \tag{3.21}$$

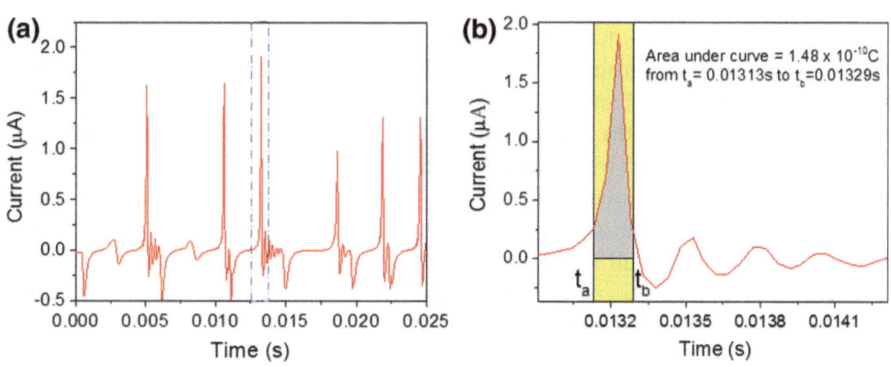

Fig. 3.14 **a** Current for TENG with a load resistance of 500 kΩ. **b** Enlarged view of current signal peak from for charge calculation

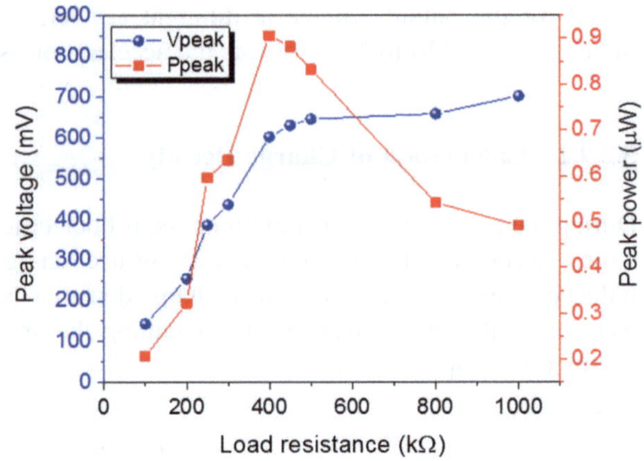

Fig. 3.15 Peak voltage and peak power characteristics generated using sample S6 at various load resistances at an acceleration of 1 g and frequency 24.5 Hz

3.2.9.3 Voltage and Power Characteristics

To calculate the power generated by cantilever TENG—I, a load resistor was connected between the top and bottom electrode. The voltage was then measured across the load resistor to obtain the power generated. As the load resistance increases, the power output increases and peaks at a point and starts dropping thereafter. Voltage output and power characteristics for sample S6 are shown in Fig. 3.15. The maximum peak power generated for sample S6 at an acceleration of 1 g was measured to be 0.91 μW and vibration excitation frequency of 24.5 Hz. The maximum power density for cantilever TENG—I was calculated to be 0.23 $\mu W/cm^2$.

3.2.9.4 Broadband Characteristics of Cantilever TENG—I

The impact between the top and bottom part introduces non-linearity in the cantilever beam spring constant. This non-linearity in the cantilever is expected to increase the operating bandwidth of resonant TENG—I device as discussed in Sect. 3.2.4. To study the broadband behavior of cantilever based TENG—I, micropillar arrays S1–S6 were tested for frequency range 10–60 Hz for acceleration levels 0.4 to 1.6 g as shown in Fig. 3.16. As the acceleration level was increased, the operating bandwidth also increased continuously. This behavior was consistently observed across all the micropillar array samples for cantilever TENG—I. The percentage changes in operating bandwidth as the excitation acceleration increased from 0.4 to 1.6 g are summarized in Table 3.3.

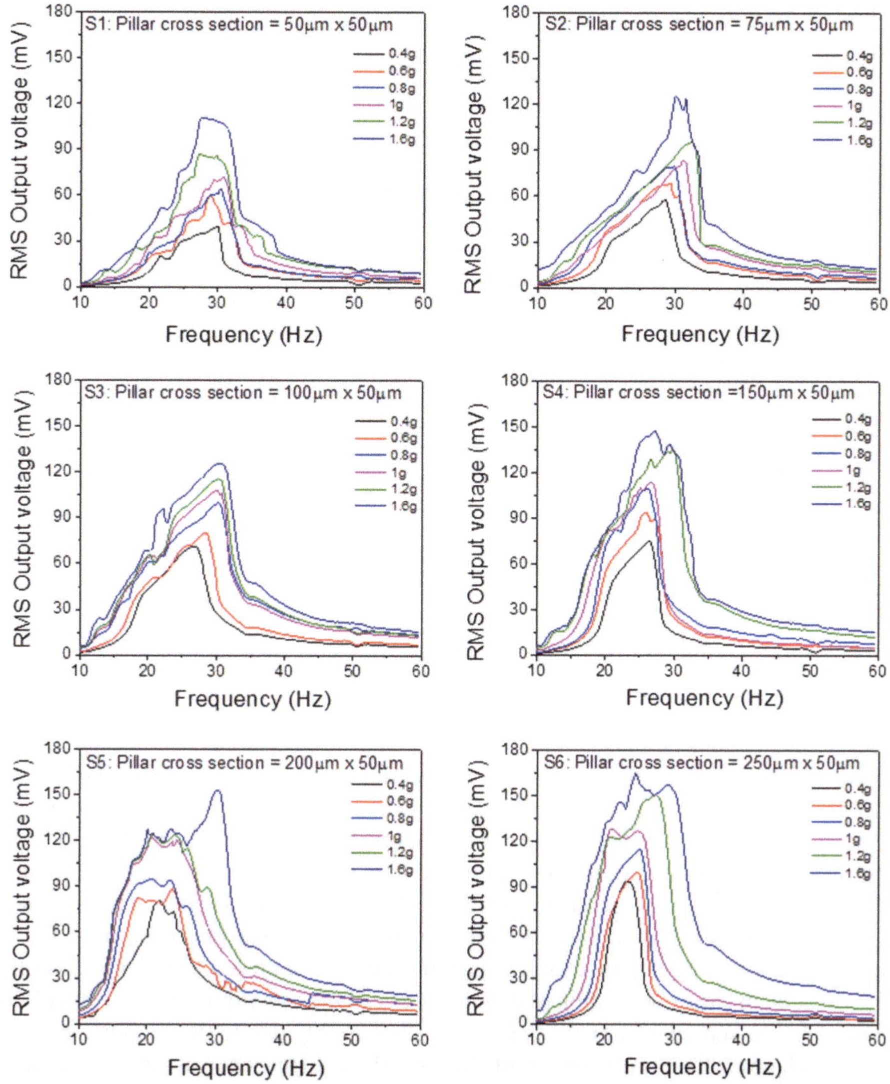

Fig. 3.16 Wideband characteristics of the voltage signal generated by samples S1 to S6

3.2.9.5 Effect of Fill Factor on Power Generation

In this work, we used rectangular micropillar shaped structures to enhance the triboelectric charging and studied the effect of different micropillar dimensions on the power generated. Experiments were conducted on micropillar arrays S1–S6. Peak power generated for different arrays is summarized in Fig. 3.17. The peak power observed can be correlated with the fill factor as defined by Eq. (3.19). As the fill

Table 3.3 Operating bandwidth

Type	RMS voltage level for bandwidth	Bandwidth at 0.4 g	Bandwidth at 1.6 g	Change in bandwidth (%)
S1	30 mV	5.36	20.5	282
S2	35 mV	7.64	21.5	181
S3	40 mV	9.43	22.05	134
S4	45 mV	6.9	17.09	148
S5	50 mV	7.05	21.3	202
S6	55 mV	4.79	18.4	284

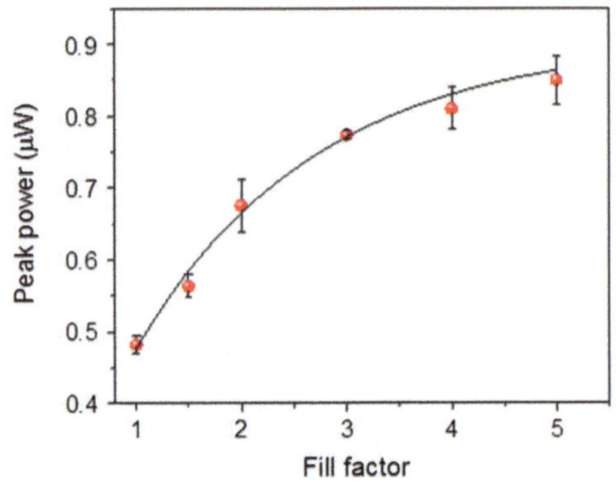

Fig. 3.17 Peak power generated for different micropillar arrays

factor is increased for the micropillar array, the peak power was observed to be increasing but has a diminishing effect as shown in Fig. 3.17. The increase in power generated is attributed to the increase in contact surface area between the PDMS layer and gold coated SU-8 micropillars. As the contact surface area increases, the contact electrification process is enhanced resulting in higher amount of triboelectric charges. But at the same time, the air voids between the micropillars also play an important role in generation of triboelectric charges. The introduction of air voids in the micropillar array results in easier separation of charges between the two triboelectric layers [2] as the two layers separate from each other. This is possibly due to two reasons. Firstly, the triboelectric effect is is significantly improved due to presence of air voids leading to higher contact area between the triboelectric layers. Secondly, in the patterned films, the triboelectric layers are easily separated leading to larger dipole moment between the electrodes [2]. Therefore the size of air voids between the micropillars affects the performance of TENG. This is also the reason that effect of increasing the micropillar dimension decreases at higher fill factors due to decreasing air void size compared to micropillar size. It can be concluded from the experimental results that increasing the micropillar dimension results in increased contact area which results in increased power output but this effect starts diminishing at higher micropillar dimensions as the air voids becomes smaller leading to lesser contact area due to local deformation of PDMS layer around SU-8 micropillars at higher value of fill factors.

3.3 Cantilever Based TENG—II

3.3.1 Device Design and Fabrication

Cantilever based TENG—II (Fig. 3.18) has a structurally similar configuration to cantilever TENG—I but the materials and microarray patterns are changed. The first triboelectric layer was made of PDMS micropad structures. The PDMS micropad array structures are highly elastic as compared to the SU-8 micropads in the cantilever TENG—I. The process flow for fabrication of PDMS micropad structures is shown in detail in Fig. 3.19a. To prepare the first triboelectric layer, silicon wafer with oxide layer was cleaned using acetone and IPA followed by oxygen plasma. After cleaning, the samples were spin coated with hexamethyldisilazane (HMDS) followed by spin coating of AZ9260. Then the sample was baked at a temperature of 110 °C for 30 min. Thereafter, the photoresist was exposed using mask to perform the photolithography. The samples were then again baked at a temperature of 120 °C for 30 min. The samples were then developed in AZ developer for 12 min. After patterning the AZ9260 photoresist, the photoresist patterns were used as mold to fabricate PDMS micropad structures. To transfer the micropad patterns on the PDMS, SYLGARD® 184 silicone elastomer kit was used. The base and the curing agent were mixed in a 10:1 ratio and the mixture was degassed in a vacuum chamber for 2 h to remove the air bubbles introduced during the mixing. After degassing, the PDMS mixture was spin coated on the silicon mold and then cured at 85 °C for 1 h. After curing the PDMS mixture, PDMS layer was peeled off to transfer micropad structures on the PDMS layer. This patterned PDMS layer was then attached to a glass slide coated with 100 nm thick layer of Cu as shown in steps in Fig. 3.19a (ii). For bonding, an uncured layer of PDMS was spin coated on the glass slide and again cured together after laying the patterned PDMS layer on it. The copper thin film on glass slide acted as the top electrode for the device. Figure 3.19b shows the optical image of the patterned PDMS layer on top of the copper thin film.

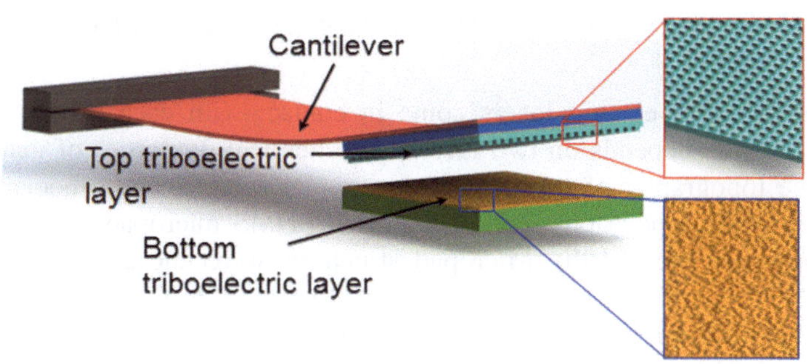

Fig. 3.18 Schematic of cantilever TENG—II

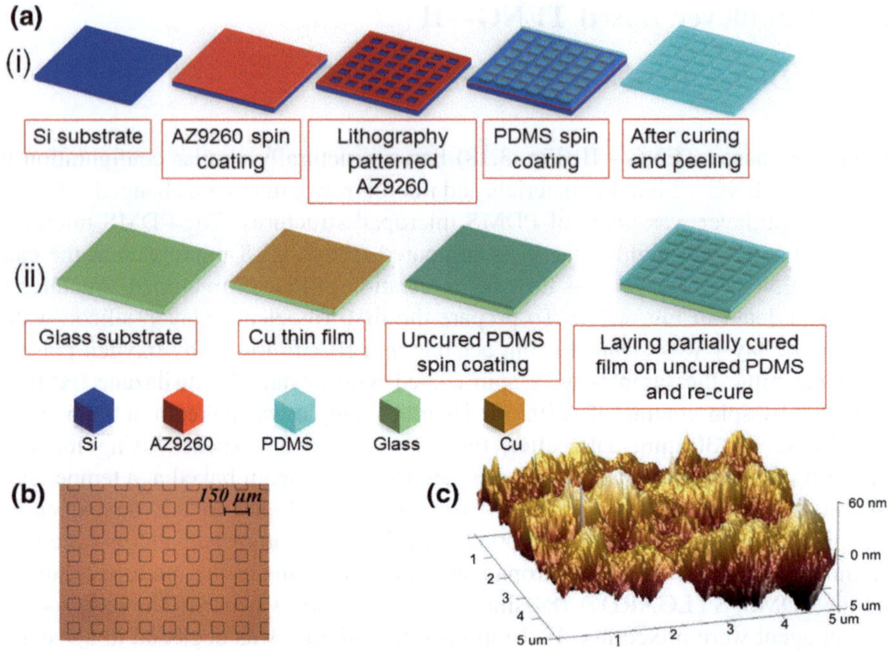

Fig. 3.19 **a** Fabrication steps for the cantilever TENG—II. **b** Optical image of the fabricated PDMS micropad structures. **c** AFM image of the plasma etched copper triboelectric layer

For the bottom triboelectric layer TENG, a 40 μm thick laminated copper layer was attached to glass substrate. The copper surface was treated with Ar/O_2 plasma at 70 W for 1 h to create nano-roughness as shown in the AFM image in Fig. 3.19c. The roughness created by plasma treatment was used to improve the performance of contact electrification during device operation.

3.3.2 Deformation in the PDMS Micropad Patterns

When the two triboelectric layers come in contact with each other, the surface charge generated depends on two factors: (i) chemical properties of materials and (ii) surface topography of triboelectric layers. To enhance the triboelectric effect using the surface topography, cuboidal shaped PDMS micropad pattern structures were fabricated (Fig. 3.19b). Micropad structures were chosen over high aspect ratio PDMS pillars because during the contact-separation mechanism required for contact electrification process, high aspect ratio PDMS pillars tend to collapse [14] which may lead to deterioration of performance of TENG after certain number of operating cycles which was to be avoided for the study.

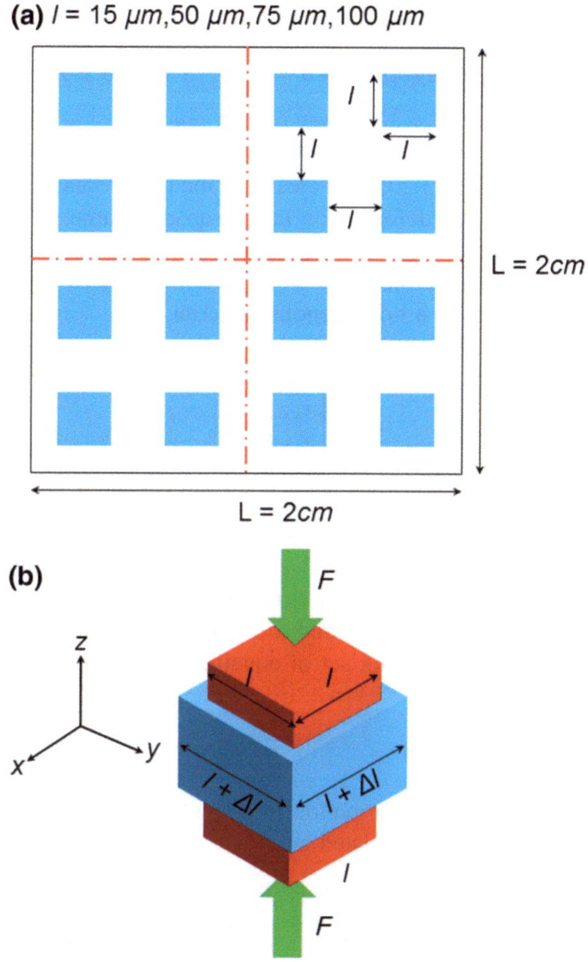

Fig. 3.20 **a** Schematic of PDMS micropad array in a square lattice. **b** Deformation in PDMS pad type structures when mechanical force is applied

Five variations of TENG devices were prepared including a plain/unpatterned PDMS layer and four different array configurations. Figure 3.20a illustrates a schematic diagram of micropad array pattern in a square lattice. The total square sample size i.e. micropad array size has a side length of L. Each individual micropad has a square cross-section of side length l. The gap between two micropads along both x-axis and y-axis is also equal to l. The array dimensions were carefully chosen such that the contact area due to the micropad top surface as a percentage of total sample size is equal for all the array patterns. This is illustrated by the following calculations:

$$\begin{aligned}&\mathit{Total\,number\,of\,micropads} = \tfrac{L}{2l} \times \tfrac{L}{2l}\\&\mathit{Total\,micropad\,area} = \tfrac{L}{2l} \times \tfrac{L}{2l} \times l^2 = 0.25L^2\\&\mathit{Total\,micropad\,area\,as\,\%\,of\,sample\,size} = \tfrac{0.25L^2}{L^2} \times 100\% = 25\%\end{aligned} \tag{3.22}$$

From Eq. (3.26), it is clear that the total contact area due to micropad top surface before elastic deformation in micropad is same for all the array configurations.

For the deformation of PDMS micropad structures, they can be modeled using generalized Hooke's law in three dimension as expressed in Eq. (3.23), assuming that PDMS is a homogenous material.

$$\begin{bmatrix}\epsilon_{xx}\\ \epsilon_{yy}\\ \epsilon_{zz}\end{bmatrix} = \begin{bmatrix}\frac{1}{E} & -\frac{v}{E} & -\frac{v}{E}\\ -\frac{v}{E} & \frac{1}{E} & -\frac{v}{E}\\ -\frac{v}{E} & -\frac{v}{E} & \frac{1}{E}\end{bmatrix}\begin{bmatrix}\sigma_{xx}\\ \sigma_{yy}\\ \sigma_{zz}\end{bmatrix} \tag{3.23}$$

where ϵ_{ii} is the strain in direction $i = x, y, z$, σ_{ii} is the normal stress in direction $i = x, y, z$, E is the Young's modulus and v is the Poisson's ratio. If we assume that in contact separation mechanism normal stress is applied in only normal z-direction on the PDMS micropad structures as shown in Fig. 3.20b, the other two normal stress terms i.e. σ_{xx} and σ_{yy} disappear from the Eq. (3.23). Therefore, from Eq. (3.23) following set of equations can be derived:

$$\epsilon_{xx} = -\frac{v}{E}\sigma_{zz}, \quad \epsilon_{yy} = -\frac{v}{E}\sigma_{zz}, \quad \epsilon_{zz} = \frac{1}{E}\sigma_{zz} \tag{3.24}$$

The elastic deformation in a single PDMS micropad structure along x-axis and y-axis can be calculated as:

$$\begin{aligned}\Delta x = l\epsilon_{xx} = -\frac{l}{E}v\sigma_{zz}\\ \Delta y = l\epsilon_{xx} = -\frac{l}{E}v\sigma_{zz}\end{aligned} \tag{3.25}$$

The change in total contact area between two triboelectric layers due to the elastic deformation in PDMS micropad structure can be calculated as:

$$\Delta A = \Delta x \times \Delta y = \frac{l^2}{E^2}v^2\sigma^2 \tag{3.26}$$

Hence, the change in total effective contact area after elastic deformation is given by:

$$n \times \Delta A = \left(\frac{L}{2l}\right)^2 \times \frac{l^2}{E^2}v^2\sigma^2 = \frac{L^2v^2\sigma^2}{4E^2} \tag{3.27}$$

where n is the total number of pillar on the PDMS triboelectric layer and is equal to $\left(\frac{L}{2l}\right)^2$. It is clear from Eq. (3.27) that the effective change in the contact area after the elastic deformation is also independent of the dimension l in micropad arrays (Fig. 3.20a).

3.3.3 *Results and Discussion*

3.3.3.1 Broadband Behavior of Cantilever TENG—II

Broadband behavior was observed in the cantilever TENG—II, due to stiffening of the cantilever spring introduced by the fixed second triboelectric layer as the first layer comes in contact. The broadband behavior can be studied by up-sweeping and down-sweeping of frequency as shown in Fig. 3.21, which shows the results for micropad array of 50 μm × 50 μm sample. An additional proof mass of 2.36 g was attached at the cantilever tip for all the experiments. When a down-sweep of the frequency was conducted, the resonant frequency of the device was observed to be ~31.5 Hz. However, when an up-sweep of frequency spectrum was conducted, broadband characteristic of cantilever TENG—II could be observed due to spring stiffening. During the up-sweep, as the top triboelectric layer attached on cantilever spring comes in contact with the bottom triboelectric layer, the stiffening occurs in the cantilever spring and hence the effective spring constant increases. Due to increase in the effective spring constant, the resonant frequency of the device increases, which further shifts the resonant peak to the right of original resonant peak. In Fig. 3.20, as the frequency increases from 31.5 to 38.8 Hz, the effective stiffness increases and the device operating frequency original range keeps broadening till point A. Thereafter, the output voltage drops abruptly [15] and both the triboelectric layers lose contact with each other for further increase in operating frequency. The operating bandwidth for excitation acceleration of 1 g at a voltage level of 70 mV RMS output voltage increased from 7.3 to 11.9 Hz, which is an

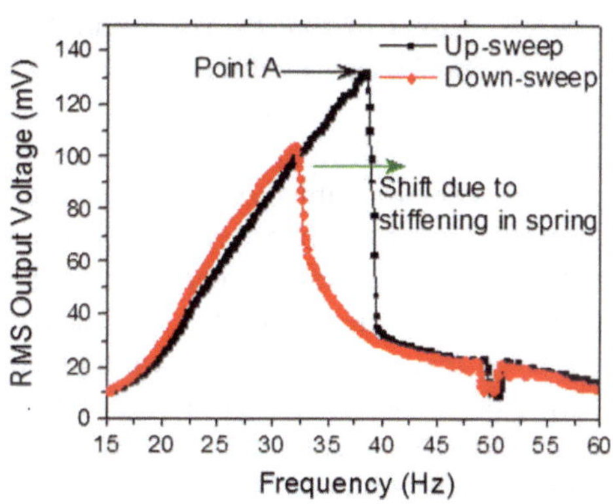

Fig. 3.21 RMS output voltage for frequency up-sweep and down-sweep at an acceleration level of 1 g for 50 μm × 50 μm sample

increment of 63% from the original operating bandwidth. It can be seen that there is a small dip in the frequency response near 50 Hz. In oscillating systems, phenomenon named ‘antiresonance’ leads to a pronounced drop in the amplitude of the oscillator at certain frequencies. This behavior is caused due to destructive interference between the external driving force and the oscillating system. The small dip observed in the frequency response is possibly due to ‘antiresonance’ phenomenon in the system during the experiment.

Furthermore, experiments were conducted to study the broadband behavior in TENG with different PDMS micropad dimensions at different excitation acceleration levels. Figure 3.22 shows the results for RMS output voltage generated by five different cantilever TENG—II devices. For all the devices, the excitation acceleration level is increased from 0.6 to 1.4 g in steps of 0.2 g. It can be observed that for all five samples that as the acceleration level is increased, the operating bandwidth also increases along with increased output level.

3.3.3.2 Output Characteristics of Cantilever TENG—II

Output voltage and current were measured for the device with micropad dimension of 50 μm × 50 μm for various values of load resistances connected. The voltage and current characteristics for 50 μm × 50 μm micropad dimensions at acceleration level of 1 g are summarized in Fig. 3.23a. The output voltage across the load resistance increases as the resistance value is increased from 50 kΩ to 11 MΩ. On the other hand, the current flowing through the load resistor decreases with the increasing load resistance value.

Figure 3.23b shows the power characteristics of cantilever TENG—II device with micropad dimension of 50 μm × 50 μm at an acceleration level of 1 g. As the load resistance is increased, the power output increases up to an optimum value of load resistance and starts decreasing thereafter. The optimum value of load resistance was observed to be 5.8 MΩ. Time domain signals of the TENG at the optimum value of resistance were also recorded as shown in Fig. 3.21c, d.

3.3.3.3 Effect of PDMS Micropad Array Configuration on Device Performance

To study the effect of PDMS micropad array configuration, five set of TENG devices were assembled on shaker and tested. Time domain results for open circuit voltages for all the five devices tested are shown in Fig. 3.24. It is interesting to notice that the overall primary contact area due to PDMS micropad cross section remains same before and after the impact as discussed earlier in Sect. 3.4.2. It was observed that although the overall contact area remains same before and after the elastic deformation in PDMS micropads, the array configurations which used the micropad cross section area divided into smaller micropads had a relatively better performance compared to bigger micropads with same overall contact area. This

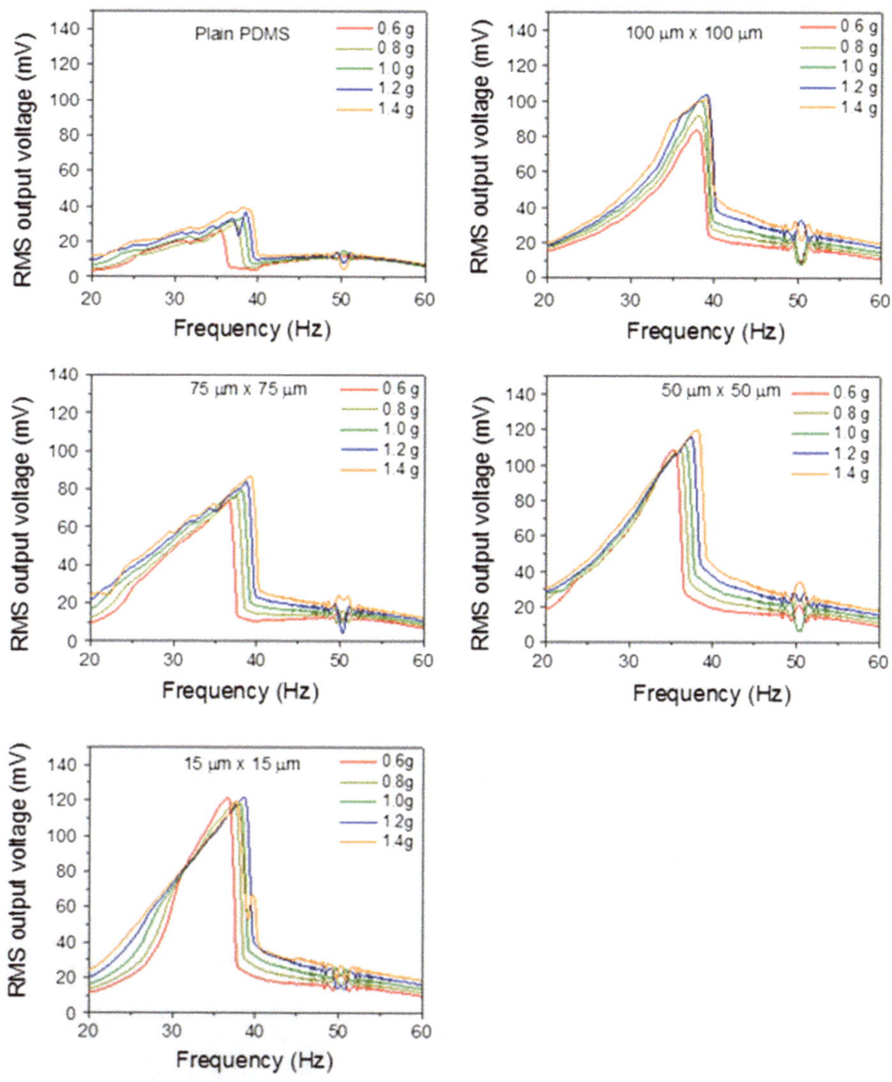

Fig. 3.22 Wideband characteristics of the voltage signal produced by different PDMS micropad samples

device behavior can be explained by the easier separation of the triboelectric charge [2] when there are sharper or smaller asperities keeping the overall contact area between the two triboelectric layers same. This result can be extended to further topographies to improve TENG performance. It is to be noted that the minimum feature size used for micropad used in the experiments was 15 μm × 15 μm. This was limited due to the process fabrication limitations. In the absence of experimental evidence, it remains debatable question that for what range of size, the

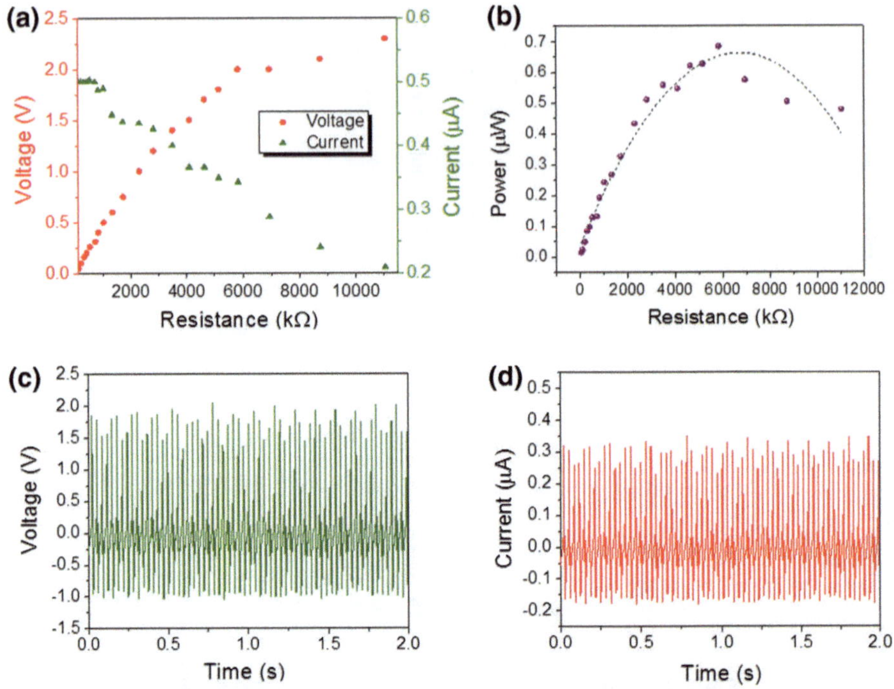

Fig. 3.23 **a** Peak voltage and current produced by cantilever TENG—II at different values of load resistances. **b** Power characteristics of cantilever TENG—II at different values of resistances. **c** Open circuit voltage and **d** short circuit current at optimal load resistance

Fig. 3.24 Comparison of time domain signals from different micropad cross-section samples

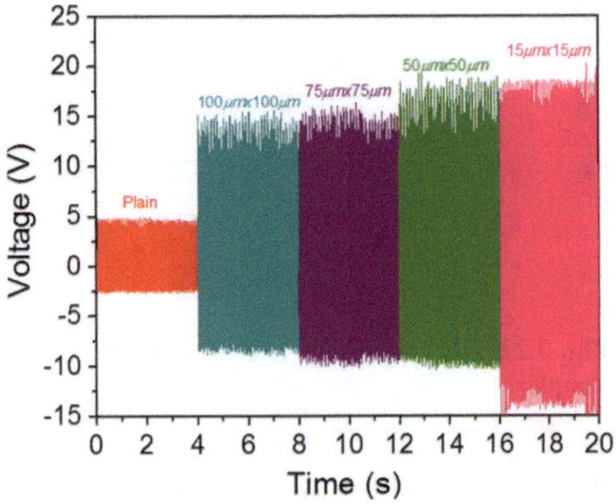

results of conducted experiments can be extended. The results provide an important guideline for structures in the range of microscale patterns, that if it is possible to pattern the triboelectric layer with similar materials with smaller dimension structures keeping the overall contact area same, the device performance of triboelectric mechanism based energy harvesters can be significantly improved.

3.4 Summary

A cantilever-based design for triboelectric mechanism based energy harvester is proposed and fabricated to capture energy from low frequency vibrations. A theoretical model was developed for cantilever mechanics and contact force between the triboelectric layers. One of the main advantages of the presented design is broadband behavior observed in the resonant TENG due to non-linearity introduced by contact-separation mechanism. The bandwidth is observed to be continuously increasing as the acceleration level is increased. The cantilever based design was utilized to understand the effect of topography on the performance of TENG.

Based on cantilever TENG—I results, power output was found to depend on the size of micropillars and air voids between micropillars. As the micropillar dimension increases the power output increases due to increase in contact area between two surfaces. But this effect starts diminishing at higher fill factors i.e. higher micropillar dimensions due to decrease in the size of air voids which results in increased difficulty in triboelectric charge separation. In conclusion, patterned triboelectric layers have better performance than unpatterned triboelectric layers and introduction of air voids with an optimum size is necessary to achieve better performance for triboelectric mechanism based energy harvesting devices [16].

Thereafter cantilever based TENG—II was used to conduct a study on the effect of size of the patterned structures on the performance of triboelectric mechanism based energy harvesting devices. It used elastic PDMS micropad structures on the performance of triboelectric energy harvesting mechanism. The performance of smaller structures was observed to be better than bigger structures, keeping the effective contact area between the triboelectric layer constant. Both the cantilever based devices were shown to demonstrate broadband characteristics [17].

References

1. C.H. Park, J.K. Park, H.S. Jeon, B.C. Chun, Triboelectric series and charging properties of plastics using the designed vertical-reciprocation charger. J. Electrostat. **66**, 578–583 (2008)
2. F.-R. Fan, L. Lin, G. Zhu, W. Wu, R. Zhang, Z.L. Wang, Transparent triboelectric nanogenerators and self-powered pressure sensors based on micropatterned plastic films. Nano Lett. **12**, 3109–3114 (2012)
3. C. Saha, T. O'donnell, N. Wang, P. McCloskey, Electromagnetic generator for harvesting energy from human motion. Sens. Actuators, A **147**, 248–253 (2008)
4. S. Roundy, P.K. Wright, J. Rabaey, A study of low level vibrations as a power source for wireless sensor nodes. Comput. Commun. **26**, 1131–1144 (2003)
5. M. Soliman, E. Abdel-Rahman, E. El-Saadany, R. Mansour, A wideband vibration-based energy harvester. J. Micromech. Microeng. **18**, 115021 (2008)
6. H. Liu, C. Lee, T. Kobayashi, C.J. Tay, C. Quan, Investigation of a MEMS piezoelectric energy harvester system with a frequency-widened-bandwidth mechanism introduced by mechanical stoppers. Smart Mater. Struct. **21**, 035005 (2012)
7. M.S. Soliman, E.M. Abdel-Rahman, E.F. El-Saadany, R.R. Mansour, A design procedure for wideband micropower generators. Microelectromech. Syst. J. **18**, 1288–1299 (2009)

8. H. Liu, C. Lee, T. Kobayashi, C.J. Tay, C. Quan, Piezoelectric MEMS-based wideband energy harvesting systems using a frequency-up-conversion cantilever stopper. Sens. Actuators A **186**, 242–248 (2012)
9. L. Dhakar, H. Liu, F. Tay, C. Lee, A new energy harvester design for high power output at low frequencies. Sens. Actuators A **199**, 344–352 (2013)
10. W. J. Stronge, *Impact mechanics*: Cambridge university press, 2004
11. S. Matsusaka, M. Ghadiri, H. Masuda, Electrification of an elastic sphere by repeated impacts on a metal plate. J. Phys. D Appl. Phys. **33**, 2311 (2000)
12. S. H. Crandall, T. J. Lardner, R. R. Archer, N. H. Cook, and N. C. Dahl, An introduction to the mechanics of solids, 1978
13. A. Narimani, M. Golnaraghi, G.N. Jazar, Frequency response of a piecewise linear vibration isolator. J. Vib. Control **10**, 1775–1794 (2004)
14. D. Chandra, S. Yang, Stability of high-aspect-ratio micropillar arrays against adhesive and capillary forces. Acc. Chem. Res. **43**, 1080–1091 (2010)
15. M.S.M. Soliman, E.M. Abdel-Rahman, E.F. El-Saadany, R.R. Mansour, A wideband vibration-based energy harvester. J. Micromech. Microeng. **18**, 115021 (2008)
16. L. Dhakar, F.E.H. Tay, C. Lee, Development of a broadband triboelectric energy harvester with SU-8 micropillars. J. Microelectromech. Syst. **24**, 91–99 (2015)
17. L. Dhakar, F. Tay, C. Lee, Investigation of contact electrification based broadband energy harvesting mechanism using elastic PDMS microstructures. J. Micromech. Microeng. **24**, 104002 (2014)

Chapter 4
Skin Based Self-powered Wearable Sensors and Nanogenerators

4.1 Motivation

Wearable electronic devices in the form of health trackers, watches and computers have made their way to the consumer market recently. But one of the major challenges with all these wearable devices is their frequent need to change or charge battery. Most of these wearable devices are directly in contact with the human skin. Therefore if skin can somehow be used as a potential way to transfer or generate energy, it would partially solve the power needs of wearable devices. This led us to think about using skin as a triboelectric layer to potentially harvest biomechanical energy. This chapter discusses application of human skin as a triboelectric layer and for sensing. A skin based triboelectric nanogenerator and a self-powered motion sensor was developed using the skin area as an active triboelectric layer. Moreover, we demonstrated a capacitance based sensor based on the capacitance change between the human skin and flexible device electrode. This sensor was later integrated with the triboelectric nanogenerator to demonstrate potential applications in self-powered sensing.

4.2 Skin Used as a Triboelectric Material

Skin is the largest organ of the human body. It has a large surface area on the human body and performs many functions such as protection, sensing and heat regulation. Due to its abundant exposed area on the human body, it can potentially be used as triboelectric layer to generate charge using contact electrification. Meng et al. [1] and Yang et al. [2], have earlier used human skin as a triboelectric layer for tactile sensing using triboelectric effect. As per triboelectric series, the outer layer of human skin also known as epidermis has a tendency to get positively charged as

L. Dhakar, *Triboelectric Devices for Power Generation and Self-Powered Sensing Applications*, Springer Theses,
DOI 10.1007/978-981-10-3815-0_4

compared to majority of polymers [3]. In this work, we propose and fabricate flexible and wearable device which use epidermis as an active triboelectric layer for energy generation and sensing.

4.2.1 Device Design

As the epidermis has a tendency to donate electrons compared to polymers, a polymer was chosen as another triboelectric layer that could be easily fabricated and patterned. We used PDMS as the polymer triboelectric layer in conjunction with epidermis as it has a high tendency to gain electrons as per triboelectric series. A schematic of the proposed device is shown in Fig. 4.1. The device comprises of PDMS nanopillars which is bonded to a polyimide layer. The polyimide layer is coated with a thin film of gold, which acts as an electrode. The device is a single electrode device, which means the charges harvested in a load resistor connected between the gold electrode and a ground reference electrode, as the relative potential at gold electrode changes due to mechanical motion. The proposed device has an advantage of simple structure and high flexibility. The fabrication of the device is discussed in the next section.

4.2.2 Device Fabrication

For the fabrication of TENG (Fig. 4.2a), an anodized aluminum oxide (AAO) substrate with nano-pores was spin-coated with a 10:1 mixture of polydimethylsiloxane (PDMS). The AAO template with PDMS layer was then kept for degassing for 2 h to allow the PDMS enter the nano-pores. Thereafter, the samples were cured at 120 °C for 2 h to cure the PDMS layer. The ~500 μm thick PDMS layer was then carefully peeled off from the AAO substrate using tweezers to transfer nanopillar structures on PDMS layer. A polyimide layer was coated with a 50 nm gold using thermal evaporation process. The gold thin film served as the electrode for TENG. The nano-patterned PDMS layer was then laid upon and bonded with the polyimide layer already coated with gold using silicone adhesive.

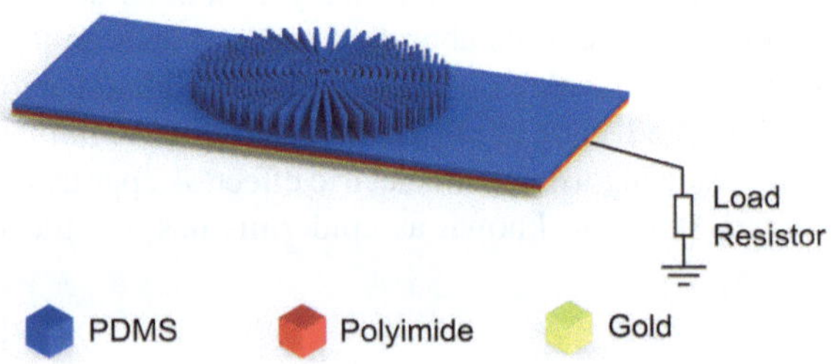

Fig. 4.1 Schematic illustration of the proposed skin based TENG

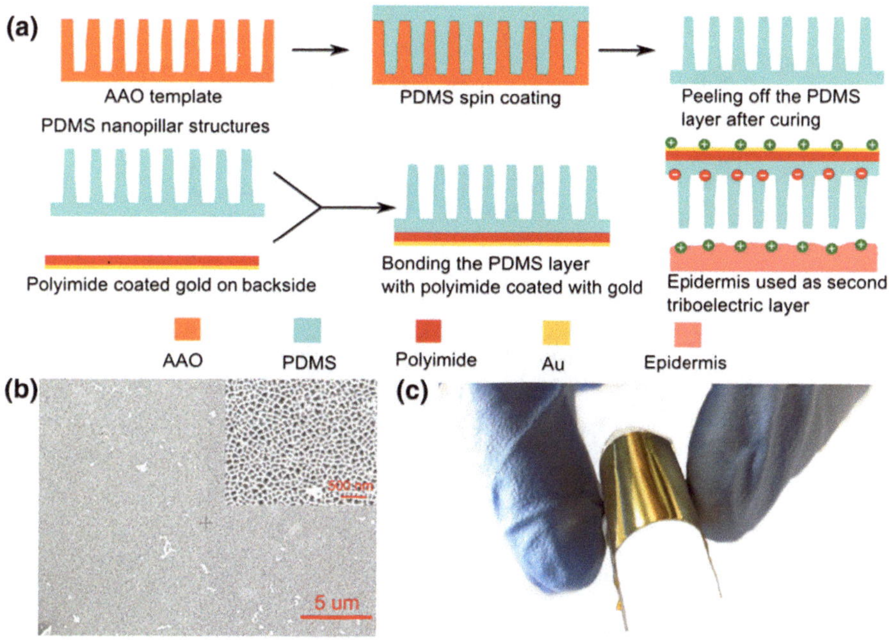

Fig. 4.2 **a** Fabrication steps for the flexible TENG. **b** Scanning electron microscopy (SEM) image of the AAO nanopores. **c** Photograph of the actual fabricated device showing the flexibility

Polyimide was chosen as the supporting layer for PDMS layer due to its mechanical properties and flexibility which helped to maintain the overall strength and robustness of the device. Figure 4.2b shows a scanning electron microscope (SEM) image of AAO template with nanopores. Figure 4.2c shows a photograph of the device depicting its flexibility. The as-fabricated device was then assembled on epidermis. The nanopillar structures increased the effective contact area between the epidermis and the PDMS triboelectric layer. During the device operation, the PDMS layer gets negatively charged whereas the human skin gets positively charged as shown in Fig. 4.2a.

The fabricated PDMS samples were characterized using Atomic Force Microscopy (AFM) in tapping mode. Figure 4.3 shows an optical image of as-fabricated PDMS sample. The sample is divided into two parts, one part is fabricated using the AAO nanopores whereas other was peeled from area without any nano-pore structures. As the device was characterized using AFM, the difference between the topography of two areas could clearly be seen. The part without AAO nanopores was flat as compared to the nano-pillar like structures on the part fabricated using AAO nanopores. This nano-structured topography results in the increased effective contact area between the epidermis and PDMS layers leading to the improved performance of device.

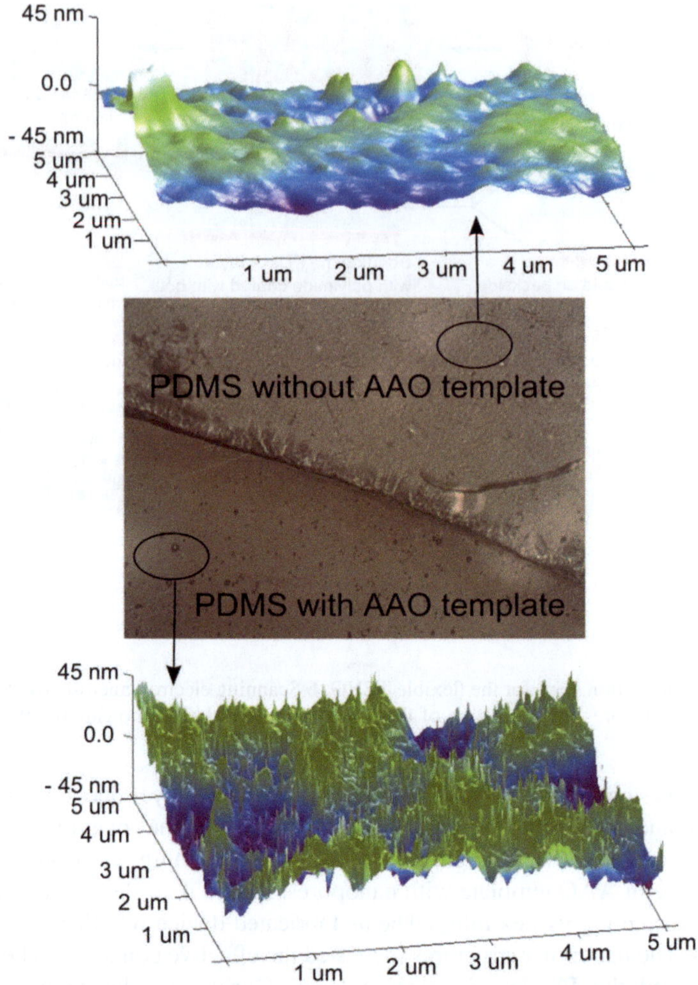

Fig. 4.3 Optical image of as-fabricated PDMS sample with AFM images of topography of parts with and without nanopillar structures using AAO template

4.2.3 Working Mechanism

The device uses PDMS and epidermis served as two triboelectric layers for surface charge generation using triboelectric effect. According to triboelectric series [4], epidermis has a high tendency to lose electrons relative to PDMS, which makes epidermis a good choice for energy generation using triboelectric mechanism. As the PDMS layer and epidermis come in full contact due to external mechanical force as shown in Fig. 4.4i, the skin gets positively charged and PDMS layer gets negatively charged. At this point, the electric field due to the charges on epidermis

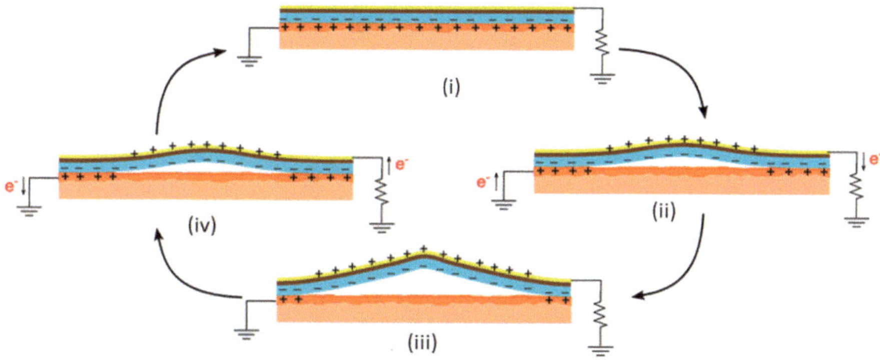

Fig. 4.4 Working mechanism of triboelectric nanogenerator utilizing skin as an active triboelectric layer for energy generation

and PDMS layer is negligible due to the close vicinity of surfaces charges generated. Thereafter, the PDMS layer detaches from epidermis as mechanical force is released. As epidermis has conductive nature and large surface area, it can be considered as ground and the charges get redistributed. After the detachment of epidermis and PDMS layer, there is a potential generated at the device electrode because of electric field generated by triboelectric surface charges. The electrode was connected to ground using a load resistor and the electrons start moving from low potential to high potential i.e. towards the ground (Fig. 4.4ii). As the device reaches the maximum separation (Fig. 4.4iii), the device is in electrostatic equilibrium. Thereafter, as the device moves in the opposite direction, the electrons start to move back towards the device electrode (Fig. 4.4iv). This energy generation cycle can be utilized to convert devices assembled on epidermis to convert mechanical energy to electrical energy.

4.2.4 Harvesting Energy Using Skin Based Triboelectric Nanogenerator from Various Human Activities

Fabricated flexible TENG was assembled on a human forearm as shown in Fig. 4.5a. The assembly step was crucial as it has to be made sure that there is a minimal gap between the assembled TENG and epidermis. As the hand is used for any activity such as lifting object or fist opening and closing, there is a movement in the muscle that translates into closure of gap and hence contact between epidermis and PDMS layer. The contact between two surfaces results in generation of triboelectric charges on both the surfaces. As the muscle device goes back to the initial position, there is an electric potential developed on the gold electrode. This electric potential changes as the gap between epidermis and PDMS layer varies. This variation of electric potential results in inflow and outflow of electrons in the

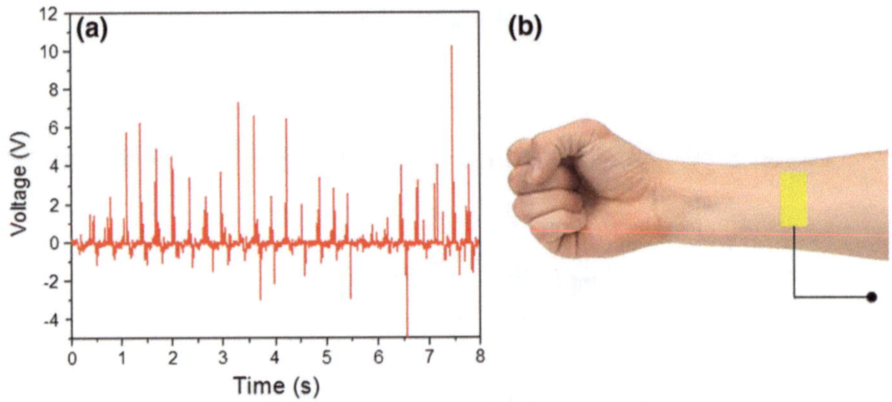

Fig. 4.5 **a** Voltage output from fist movement of hand. **b** Skin based nanogenerator assembled on human wrist

load resistor connected between the electrode and ground. This electron flow can be converted into useful electrical energy in the load resistor.

To measure the voltage output response of TENG, first set of experiments were conducted using with fist clenching and releasing activity of human hand. As the fist is clenched, there is a movement in the muscle resulting into the contact between epidermis and PDMS layer. As the muscle moves back to the initial position, the two triboelectric layers separate from each other. This contact-separation cycle generated a peak voltage up to 10 V.

Furthermore, the device was also used to harvest energy from the movement of vocal box during speaking as shown in Fig. 4.6a. The device was assembled on the throat and generated a peak voltage of up to 0.25 V as different alphabets are spoken as marked in Fig. 4.6a. The device generated a peak voltage of up to 17 V from the muscle motion caused due to finger snapping, when assembled on the forearm. All these measurements were conducted using an oscilloscope with an internal impedance of 1 MΩ.

To demonstrate the ability of TENG to power electronic components as a power source, the device was directly connected to 12 commercially available LEDs connected in series. The device was again assembled on human skin and tapped mildly using finger to generate energy. As the device was touched and released with the finger, the LEDs were lighted up using the power generated by TENG. The highest peak voltage generated by the finger tapping on TENG assembled on epidermis was measured to be 90 V as shown in Fig. 4.7a. Figure 4.7b shows an image of the lit up LEDs.

These set of experiments demonstrated that TENG can potentially be used for powering small electronic devices by harvesting biomechanical energy. TENG can also be assembled on other body parts e.g. leg, arm etc. where the muscle

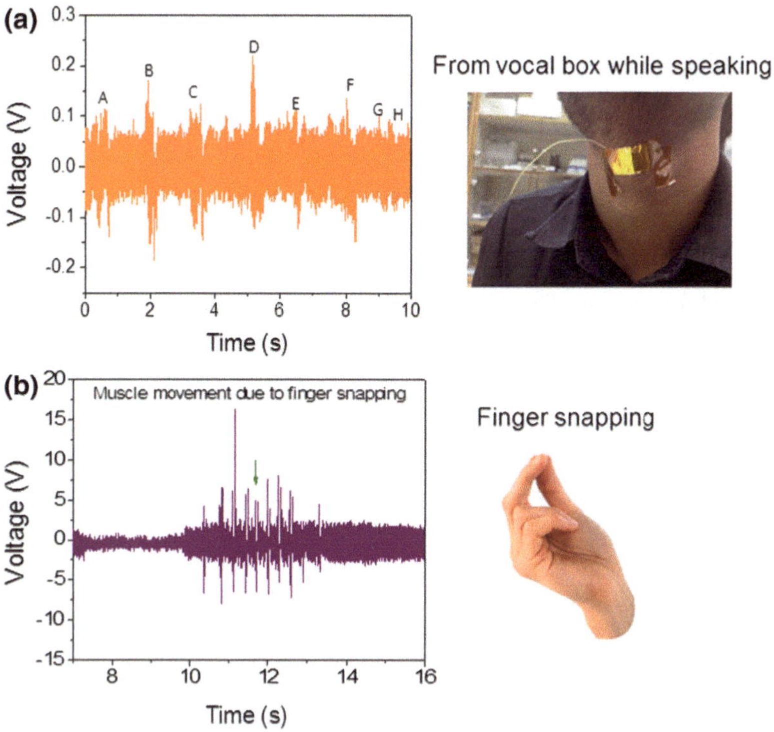

Fig. 4.6 **a** Voltage signal generated by speaking different alphabets when flexible TENG is assembled on the vocal box on throat. **b** Voltage signal generated by the finger snapping motion and the device assembled on forearm

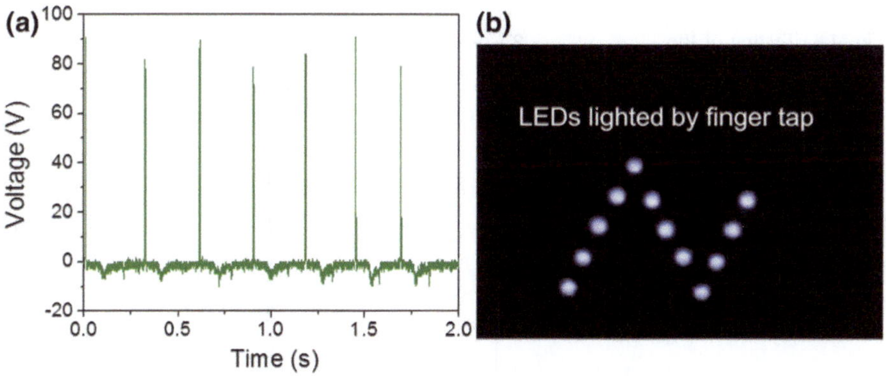

Fig. 4.7 **a** Voltage output from finger tapping. **b** LEDs lighted using mild finger touch on skin based TENG

movement due to daily activity like walking, running, lifting objects, can enable the contact and separation movement between the epidermis and nano-structured PDMS layer.

4.2.5 Testing as a Motion Sensor

Alternatively, the electrical output signal generated from the flexible TENG can also be used for sensing. The flexible TENG device was demonstrated to be used as wearable sensor. The device was assembled on hand and tested as a sensor for muscle movement due to any motion or activity of hand. As seen in Fig. 4.8, a pulse is generated due to holding of an object using hand. The object was a ball with a diameter of 7 cm and weight of 26 g which was held using the hands leading to the pulse generated due to muscle motion. The signal to noise (SNR) ratio for the TENG as a motion/activity sensor can be calculated using Eq. (4.1).

$$SNR = \left(\frac{A_{signal}}{A_{noise}}\right)^2 \tag{4.1}$$

where A_{signal} is the amplitude of required signal and A_{noise} is the amplitude of noise. As calculated from the acquired signal A_{signal} and A_{noise} were measured to be 6.34 V whereas and 0.19 V, respectively. Using these values in Eq. (4.1), the SNR is calculated to be 1113.5. The calculated SNR value demonstrates the superior characteristics of TENG as a motion or activity sensor.

This skin based wearable sensor can be used for fall detection in elderly, soft robotics and flexible electronics applications.

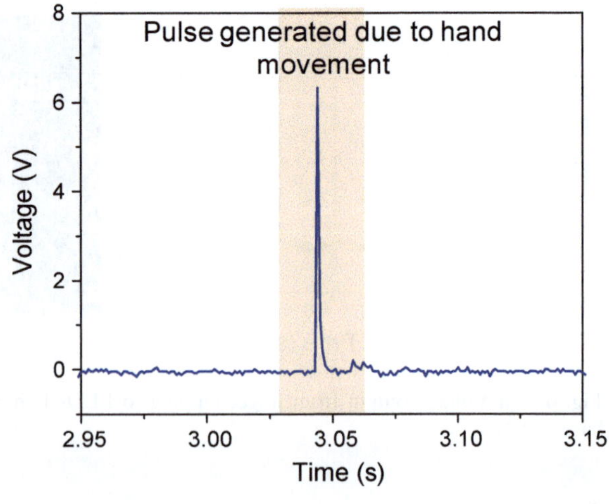

Fig. 4.8 Testing of the flexible TENG as a skin based motion/activity sensor

4.3 Integration of Skin Based Nanogenerator with a Capacitance Based Sensor to Realize Human Finger Motion Tracking

The motion sensor described in previous section can only be used for applications to detect dynamic motion as it requires the device to periodically contact and separate from epidermis to generate an electrical signal. To improve the usability of the motion sensors, it is important to develop sensors which can provide electrical signal as output data for any given or static position or motion profile. The electrical output data can then readily be calibrated with different motion parameters. In this section, a novel sensing mechanism is proposed which can be used to detect both static and dynamic finger joint positions and can readily be integrated with skin based nanogenerator as discussed in earlier sections in this chapter.

Humans are always surrounded by low frequency electric fields which are generated due to alternating current (AC) carrying power supply lines and electrical household equipment like monitors, electric oven, refrigerators etc. The effect of these low frequency electric fields on human body are well studied and are proven to be completely harmless within specified limits [5]. The electric field sensing has been utilized for sensors and studied previously. It was first proposed by Zimmerman et al. [6] at MIT Media Lab, for proximity detection and human-computer interfacing to develop personal area networks (PAN). Here, an electric field based mechanism is proposed and demonstrated to realize flexible and wearable human motion sensors. The current carrying conductors (or power lines) are capacitively coupled to epidermis [7] due to epidermis' conductive nature. The fabricated flexible device is assembled on epidermis which in turn leads to capacitive coupling between epidermis and device electrode. Due to the afore-mentioned arrangement, there is an oscillating potential signal observed at device electrode. The variation in intensity of this oscillating potential signal due to change in capacitive coupling between epidermis and flexible gold electrode was detected as a measured signal for finger motion. A proof of concept device is demonstrated to detect the motion of human finger using the proposed mechanism.

For the proposed device, energy harvesting capability is demonstrated based on triboelectric mechanism using similar device structure as discussed in Sect. 4.2.1. The harvested biomechanical energy can be used to power auxiliary electronic circuitry for the sensor and acts as a sustainable source of power. The signal characteristics due to capacitive coupling with current carrying conductor and energy generation due to triboelectric phenomenon can easily be differentiated in terms of amplitude and frequency characteristics. The signal due to capacitive coupling of low frequency electric fields is typically very low as compared to triboelectric output but is strong enough to be detected. The device is shown to generate a maximum voltage of 70 V and a current of 2.7 $\mu A/cm^2$ at a load resistance of 5 $M\Omega$. The proposed mechanism is suitable for skin and body based wearable and flexible sensors. These sensors can be used for various applications in recognition of gesture and motion pattern, advanced input method for human-machine interface and osteoarthritis rehabilitation.

4.3.1 Fabrication

The device structure of the proposed sensor is similar to the device discussed in Sect. 4.2.1. The only difference is in the topography of PDMS layer was replaced by micro-pyramid patterns instead of nanopillars. The advantage of the micro-pyramid patterns is that they could be fabricated with high uniformity and repeatability as compared to nanopillars. For the device fabrication, 100 nm thick silicon nitride (SiN) was deposited using plasma enhanced chemical vapor deposition (PECVD) on a silicon (Si) wafer. Square window opening patterns of 2 μm × 2 μm size were then etched in SiN layer using reactive ion etching (RIE). The SiN layer acted as a hard mask and was used to etch Si anisotropically leading to pyramid shaped patterns using potassium hydroxide (KOH) solution (Fig. 4.9a, c). The etched wafer was then treated with (tridecafluoro-1,1,2,2-tetrahydrooctyl)-1-trichlorosilane (United Chemical Technologies) to reduce the adhesion of polydimethylsiloxane (PDMS) with wafer during the peeling-off process. Thereafter, the patterned wafer was used as a mold and poured with 10:1 PDMS mixture and degassed in a desiccator for 4 h to let PDMS enter in KOH etched pyramid cavities. The samples were then cured at 120 °C for 1 h and the micro-patterned PDMS films were peeled off from the mold (Fig. 4.9d). These micro-patterned films were then carefully laid and bonded on a polyimide substrate which was already coated with a 50 nm gold film on the other side as an electrode. Polyimide was chosen as the substrate, owing to its excellent adhesion properties with gold film compared to PDMS and its superior mechanical characteristics. Polyimide layer also helped to maintain the overall flexibility of device while improving the mechanical robustness.

The topography of surfaces was characterized using Nova NANOSEM 230 from FEI. The voltage and current measurements are made using DSOX3034A oscilloscope from Agilent Technologies. For the capacitance measurement Agilent 4924A Precision Impedance Analyzer was used. Agilent 33500B waveform generator was used to generate AC signal at 1 kHz.

4.3.2 Operating Principle of Sensor

Humans are constantly exposed to low frequency electric fields in their daily lives. The AC carrying conductors and electrical equipment are capacitively coupled to epidermis due to its conductive characteristics [8]. A schematic diagram depicting electric field lines coupled to human body is shown in Fig. 4.10a. Epidermis can be considered as a large conductor capacitively coupled to current carrying conductor. The flexible device comprising of micro-patterned PDMS film at the top and a gold

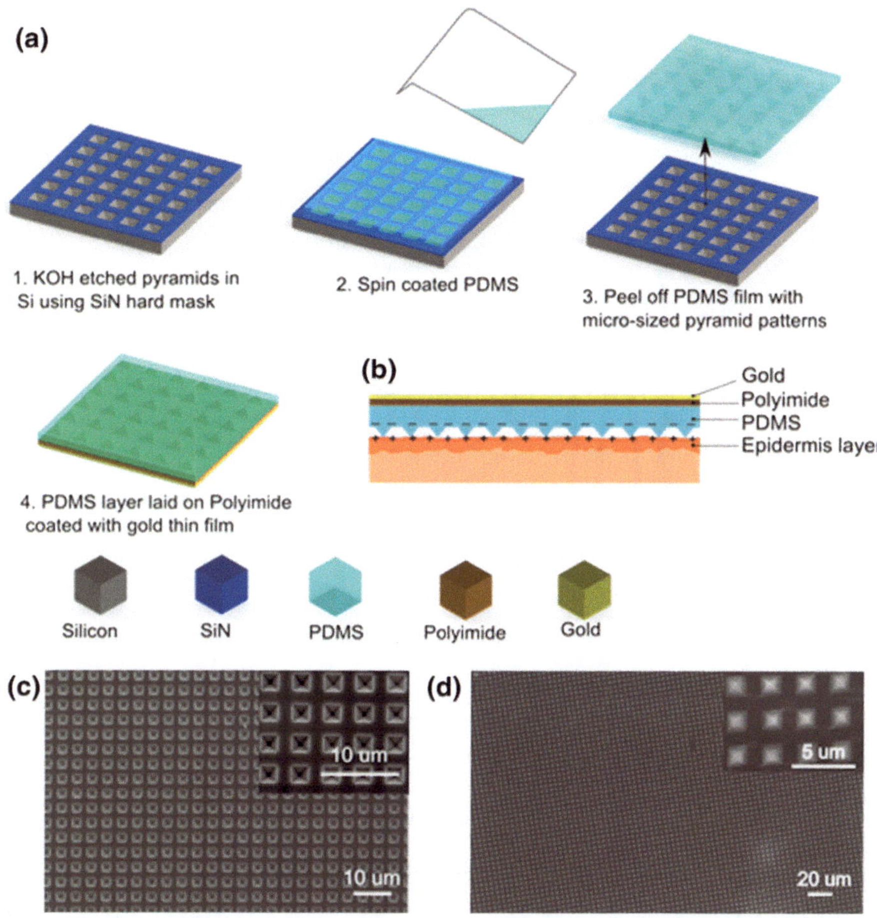

Fig. 4.9 **a** Fabrication process flow of flexible device. **b** Schematic of assembled device on epidermis layer. **c** SEM images of pyramids etched pits in silicon wafer and **d** pyramid structures transferred to PDMS

electrode at the bottom is attached to epidermis such that PDMS film is in direct contact with epidermis layer (Fig. 4.10a inset). The space between epidermis and gold electrode is separated by a combination of air, PDMS and polyimide layers. This system of stacked layers forms a capacitive system between epidermis and gold electrode. An equivalent electrical circuit is shown in Fig. 4.10b for this system. In this circuit, V_A is the potential at current carrying conductor or power line which generates electric field. The epidermis is capacitively coupled to current carrying conductor and ground conductor through capacitor C_{B1} and C_{B2}, respectively. As the device is assembled on epidermis, it is coupled to epidermis through capacitance C_{FD} which is shown as a variable capacitance in the electrical circuit. The change in capacitive coupling between epidermis and gold electrode (C_{FD})

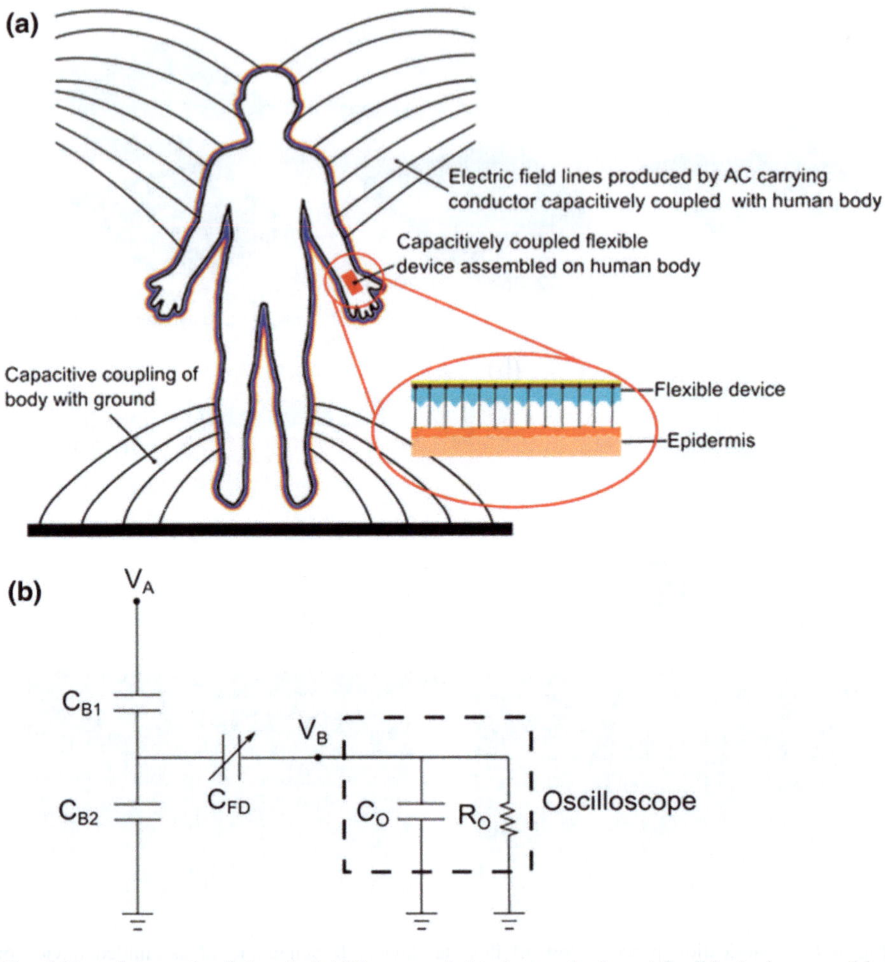

Fig. 4.10 **a** Schematic of electric field lines coupled to human body and flexible device assembled on epidermis **b** circuit diagram to model the capacitive couplings of human body and assembled flexible device on epidermis

results in the change in intensity of oscillating potential signal at gold electrode. In the presented device, we correlate the change of finger joint angle, which is the physical parameter to be measured with the change in C_{FD}, using variation in intensity of oscillating potential signal. The capacitive coupling between gold electrode and conductor producing electric field lines can safely be ignored in this model as compared to C_{B1} due to its small area relative to the overall epidermis area. The oscillating potential signal from the flexible gold electrode is then measured using oscilloscope as shown in Fig. 4.10b. C_o and R_o are used to model the oscilloscope internal capacitance and resistance, respectively. This electrical circuit was used to model and study the signal V_B using LTspice for various values of C_{FD}.

Table 4.1 Parameters for circuit simulation

Parameter	Value
C_{B1}	1 pF
C_{B2}	1 pF
C_0	8 pF
R_o	10 MΩ

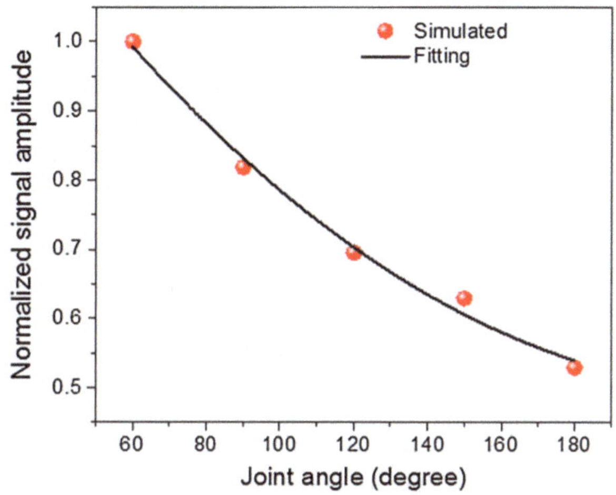

Fig. 4.11 Signal amplitude simulated using LTspice as the capacitive coupling between epidermis and device electrode changes

Model described in Fig. 4.10b is used to simulate the signal acquired by the device as the capacitive coupling changes due the index finger movement. The values used to conduct the simulations using LTspice are summarized in Table 4.1. The simulated results are shown in Fig. 4.11.

The value of C_{FD} used for simulation were 6.9 pF, 3.6 pF, 2.4 pF, 2 pF and 1.2 pF for 60°, 90°, 120°, 150° and 180° respectively, as measured in the experiments as shown in Fig. 4.13. V_A was assumed to be 220 V, 50 Hz for the simulation.

4.3.3 *Working of Triboelectric Nanogenerator*

Apart from acting as a finger motion sensor, the device also serves dual purpose of harvesting biomechanical energy from the finger movement. It worked as a single electrode triboelectric nanogenerator and works on the same principle as discussed in Sect. 4.2.3. It used PDMS micro-pyramid patterns as one of the triboelectric layers to improve the contact area when two triboelectric layers are put in contact [9]. Epidermis served as the second triboelectric layer for charge generation using triboelectric mechanism during device operation. This device configuration is also aligned with the working mechanism of motion sensor described earlier, as PDMS and epidermis are in direct contact with each other. The flexible device was attached

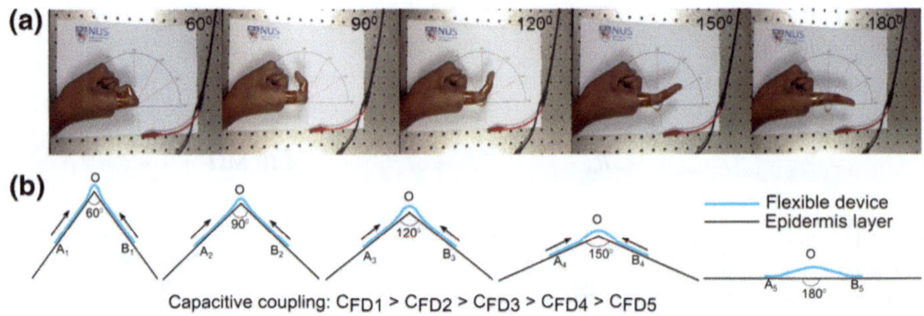

Fig. 4.12 **a** Flexible device assembled on index finger to measure the angle at finger joint starting from 60° to 180°. **b** Schematic illustration of change in capacitive coupling C_{FD} between epidermis and flexible device electrode as the finger joint angle changes

on the finger such that any movement in finger motion led to an increase in gap between the PDMS layer and epidermis (Fig. 4.12a, b). When the angle at finger joint is 60°, PDMS and epidermis are in full contact. As the joint angle increases due to finger movement, the flexible device detaches from the epidermis and electrons flow towards the ground. As the device reaches the maximum position when finger joint angle is 180°, electrostatic equilibrium is reached. Thereafter, as the finger moves in the opposite direction, the electrons start to move back towards the gold electrode. As the finger completes one full cycle of 60°–180°–60°, one cycle of power generation is completed. This kind of natural motion is generated in day to day activities when we are holding objects, typing on keyboard, fist clenching and releasing etc. More importantly, such natural finger motion can be utilized to generate energy for useful purposes.

The two features of proposed device: capacitive coupling change for motion sensing and triboelectric energy generation cycle, utilize same device configuration when assembled on epidermis. Both mechanisms work in tandem for sensing and energy generation. The finger joint motion sensor has potential applications in advanced input devices to recognize gestures and human computer interfaces. Another important application for joint motion sensing is in rehabilitation of osteoarthritis patients who suffer from difficulty in movement of finger joints. The harvested energy using triboelectrification can be used to power auxiliary electronics needed for sensor to realize a fully independent self-powered wearable sensors.

4.3.4 Finger Motion Sensor Testing

The as-fabricated device was assembled on index finger to measure the angle at finger joint as shown in Fig. 4.12a. The device was assembled such that patterned PDMS layer and epidermis were in direct contact when finger was fully bent (at angle of 60°). The range of the movement of middle joint (proximal interphalangeal joint) of index finger was measured to vary from a minimum of 60° to a maximum of 180° (angle at the joint). The schematic of the flexible device assembled on epidermis layer is shown in Fig. 4.12b. Point O is located at the joint whereas points A_i and B_i (i = 0–5, where i depicts the position of finger at an angle of 60°, 90°, 120°, 150° and 180°, respectively) are located at the end points of the flexible device where it is attached to epidermis. The epidermis is flexible and is in fully stretched position at 60° and reaches relaxed position as the joint angle increases, leading to wrinkling of the skin (epidermis) at an angle of 180°. The flexible device reaches a slack state as the joint angle increases and epidermis reaches a relaxed position. As shown in schematic diagram in Fig. 4.12b, the points A_i and B_i starts moving towards O as joint angle increases and epidermis approaches a relaxed position. The decrement in distance OA_i ($OA_1 > OA_2 > OA_3 > OA_4 > OA_5$) leads to increased gap between the PDMS film and epidermis. As the flexible device is assembled on epidermis, they form a capacitive system between epidermis and gold electrode. For simplicity, assuming there is only air separating the electrode and epidermis, the capacitance C_{FD} can be obtained using equivalent capacitance equation for capacitance connected in parallel configuration (Eq. 4.2).

$$C_{FD} = \int_{A_i}^{B_i} \varepsilon_0 \frac{w}{d} \, .dl \tag{4.2}$$

where ε_0 is the electric constant, w is the width of flexible device, d is the gap between electrode and epidermis, and varies across the epidermis surface and 'dl' is infinitesimal length across the epidermis surface. It is clear from the equation, that as the average gap d increases, it will lead to increase in capacitance $C_{FD.}$ Measurements were conducted for capacitance C_{FD} at different joint angles as shown in Fig. 4.13a. The measurements were taken at five different angles: 60°, 90°, 120°, 150° and 180°. For the capacitance measurements at different joint angles, it was found that the capacitance decreases as the joint angle increases (average gap d increases) as predicted by Eq. (4.2). The capacitance varied from 6.9 pF at 60° to 1.2 pF at 180°. From the experimental measurements, it is clear that increase in joint angle results in decrease in capacitive coupling between flexible gold electrode and epidermis. Figure 4.13b shows the change in capacitance as the finger joint angle changes. The measurements were also conducted for dynamic movements of finger moving in cycle from 60°–180°–60° as shown in Fig. 4.13c. The capacitance changes cyclically as the joint angle changes. The sensitivity using capacitance ($\delta(\Delta C_{FD}/C_{FD0})/\delta\theta$) is plotted in Fig. 4.13d which is defined by change in capacitance C_{FD} per unit change in the joint angle θ. The sensitivity was observed to increase with the increasing joint angle.

After capacitive coupling measurements between the epidermis and flexible gold electrode, measurements were conducted for oscillating potential signal due to electric field in environment. We conducted measurements using oscilloscope probes to observe the change in variation of oscillating potential signal amplitude at gold electrode as the capacitive coupling C_{FD} changes. As finger joint angle decreases, the capacitance between the epidermis and gold electrode decreases

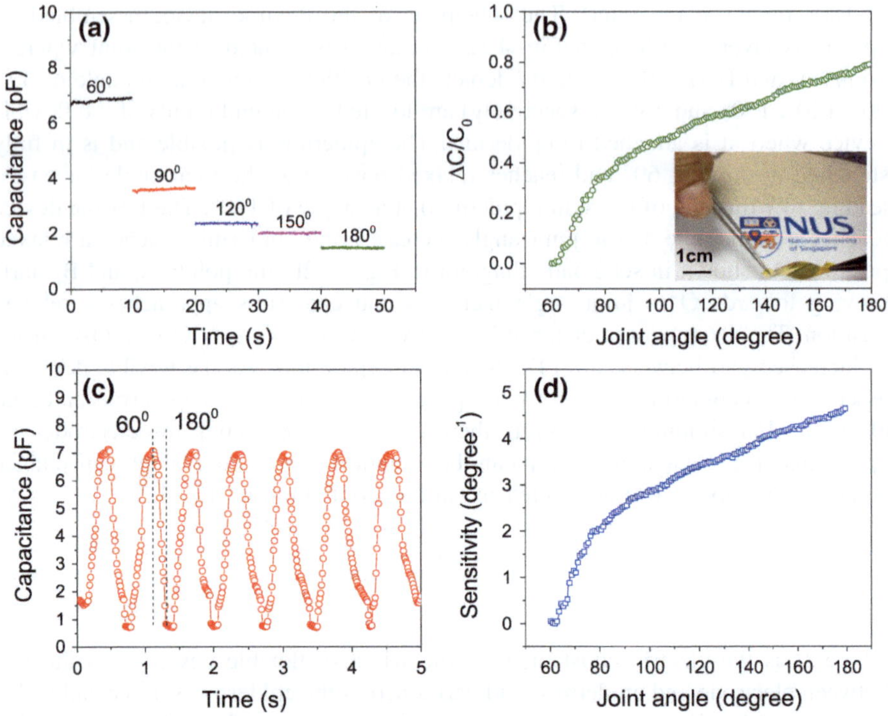

Fig. 4.13 **a** Measurement of capacitance between epidermis and flexible device electrode (C_{FD}) at different finger joint angles. **b** Change in capacitance (C_{FD}) as a function of finger joint angle, optical image of the flexible device (*inset*). **c** Time variation of capacitance as the finger joint is moved in cycles from 180°–60°–180°. **d** Sensitivity in terms of change in capacitance per unit joint angle ($\delta(\Delta C_{FD}/C_{FD0})/\delta\theta$)

leading to reduction in capacitive coupling. The oscillating potential signal at electrode is expected to decrease as the coupling between epidermis and gold electrode decreases (simulation results shown in Fig. 4.11). It was observed that the acquired signal amplitude decreases as the joint angle decreases as shown in Fig. 4.14a. Normalized signal amplitude values are plotted in Fig. 4.14a for different joint angle values. The normalized signal amplitude continuously decreases as the finger joint angle decreases. All these measurements were conducted for 50 Hz signal due to the electric field created by AC carrying conductors in our lab. Special care was taken during the experiments such that hand did not change the position while conducting a set of experiments for 50 Hz oscillating electric field. To further confirm the working principle of motion sensor, 1 kHz signal was applied to epidermis using a signal generator. The measured signal was processed by passing through a high pass filter of 100 Hz to remove the signal due to 50 Hz electric field. The results for 1 kHz signal are shown in Fig. 4.15. The oscillating potential signal measured for 1 kHz followed the same pattern as measured for 50 Hz signal measurements and confirmed the working principle theory.

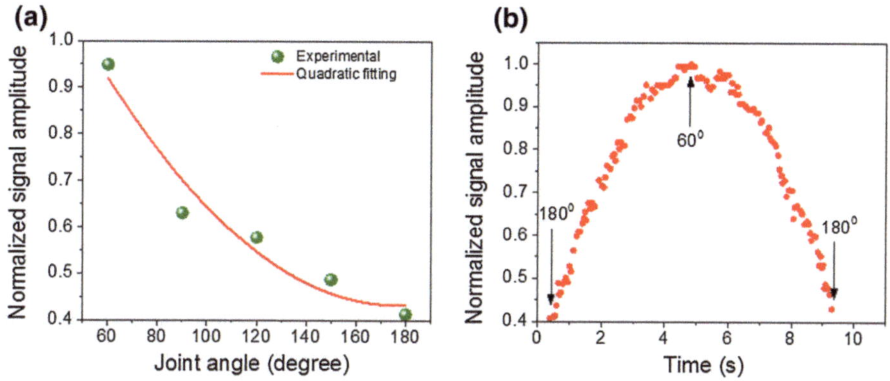

Fig. 4.14 **a** Plot of normalized signal amplitude acquired by flexible device electrode as a function of finger joint angle. **b** Variation of normalized signal amplitude as the finger joint angle completes a cycle from 180°–60°–180°

Fig. 4.15 Signal amplitude measured for 1 kHz signal capacitively coupled to skin

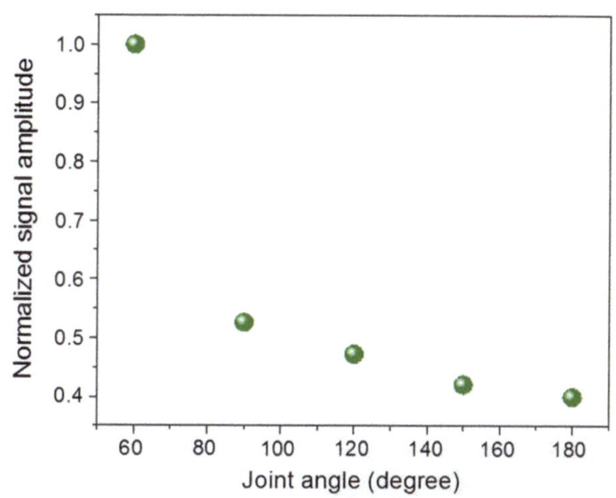

The measurements were also conducted for dynamic movement of finger as shown in Fig. 4.14b. The normalized signal amplitude increases from 0.4 at 180° and reached a maximum value of 1 at finger joint angle of 60° and then again drops down as joint angle becomes 180°. The average angular velocity of finger for the motion was calculated to be 0.42 rad/s. The angle data gathered though sensor was used to simulate the finger movement using MATLAB. The equations used to simulate the motion are given in Appendix. One of the limitations of these kind of sensors is that they are highly dependent on the coupling between epidermis and AC carrying conductor which is very sensitive to the position of body in the electric field mesh in the environment. This limitation can be overcome in future by having a controlled electric field at a specific frequency which is dedicated to human motion sensing. This can also be used in robotics for electronic skin applications.

4.3.5 Energy Harvesting Testing

For the energy harvesting testing, the flexible device assembled on index finger was used to harvest biomechanical energy from the moving motion of finger. Figure 4.16a shows the voltage generated by the flexible device utilizing biomechanical energy harvesting from finger motion. The peak voltage generated was measured to be about 70 V with a load resistance of 5 MΩ connected in parallel with the device. Figure 4.16b depicts the current area density generated by the flexible nanogenerator. The peak current area density was measured to be 2.7 $\mu A/cm^2$. The power generated by the flexible nanogenerator using finger joint movement was used to power commercially available LEDs as shown in Fig. 4.16c. First part of Fig. 4.16c shows the LEDs in OFF positions and second part shows the LEDs are lighted up using the biomechanical energy harvested by device using triboelectric mechanism. This demonstrates the capability of flexible nanogenerator to power small electronic devices or electronic circuitry from simple finger movement.

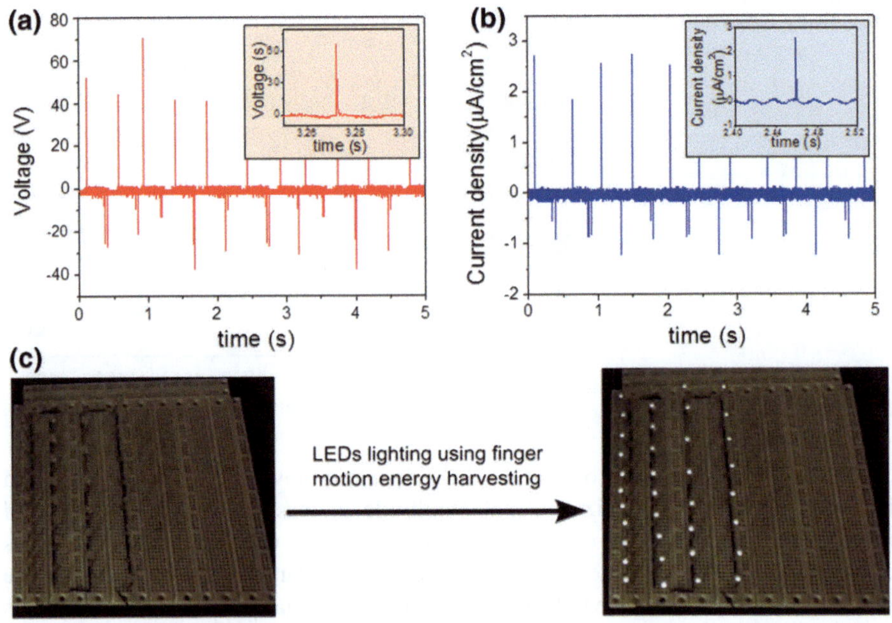

Fig. 4.16 Tribolelectric energy generation characteristics of flexible device **a** voltage generated **b** current generated by flexible device due to the index finger movement in full cycles at a load resistance of 5 MΩ **c** demonstration of LED lighting using the energy generated by flexible device assembled on index finger

4.4 Summary

A skin based flexible energy harvester is proposed and developed. The first set of devices fabricated using PDMS nanopillars were demonstrated to harvest energy using different human activities like holding objects, speaking and finger snapping. The device was also demonstrated as a motion sensor.

A new sensor mechanism is demonstrated to measure the angle at the joints for different positions of finger statically and dynamically. It can also utilize the same device structure to harvest biomechanical energy using triboelectric effect. It uses epidermis as an active triboelectric layer for charge generation which can be used for energy generation to power body wearable sensors using only one fabricated flexible triboelectric layer. The proposed sensing mechanism is very suitable for human body based intelligent wearable sensors and can be utilized in future to realize completely independent self-powered sensors.

References

1. B. Meng, W. Tang, Z.-H. Too, X. Zhang, M. Han, W. Liu et al., A transparent single-friction-surface triboelectric generator and self-powered touch sensor. Energy Environ. Sci. **6**, 3235–3240 (2013)
2. Y. Yang, H. Zhang, Z.-H. Lin, Y.S. Zhou, Q. Jing, Y. Su et al., Human skin based triboelectric nanogenerators for harvesting biomechanical energy and as self-powered active tactile sensor system. ACS Nano **7**, 9213–9222 (2013)
3. J.H. Lee, R. Hinchet, T.Y. Kim, H. Ryu, W. Seung, H.J. Yoon et al., Control of skin potential by triboelectrification with ferroelectric polymers. Adv. Mater. **27**, 5553–5558 (2015)
4. P.E. Shaw, Experiments on tribo-electricity. I. The tribo-electric series. Proc. R. Soc. Lond. Series A, Containing Papers Math. Phys. Character **94**, 16–33 (1917)
5. M.A. Stuchly, T.W. Dawson, Interaction of low-frequency electric and magnetic fields with the human body. Proc. IEEE **88**, 643–664 (2000)
6. T.G. Zimmerman, Personal area networks: near-field intrabody communication. IBM Syst. J. **35**, 609–617 (1996)
7. T.G. Zimmerman, J.R. Smith, J.A. Paradiso, D. Allport, N. Gershenfeld, Applying Electric Field Sensing to Human-Computer Interfaces, in *Proceedings of the SIGCHI Conference on Human Factors in Computing Systems*, 1995, pp. 280–287
8. D.T. Lykken, P.H. Venables, Direct measurement of skin conductance: a proposal for standardization. Psychophysiology **8**, 656–672 (1971)
9. L. Dhakar, F. Tay, C. Lee, Investigation of contact electrification based broadband energy harvesting mechanism using elastic PDMS microstructures. J. Micromech. Microeng. **24**, 104002 (2014)

Chapter 5
Large Scale Fabrication of Triboelectric Energy Harvesting and Sensing Applications

5.1 Motivation

One of the important factors affecting the performance of tribo-electrification process is the surface topography of the contact surfaces. It plays a pivotal role in the charge generation process during contact electrification [1–3]. To improve the surface topography and properties, different methods have been used including lithography patterning [4, 5], plasma etching [6–9], nanoparticle deposition [10], block polymer self-assembly [11, 12] and chemical treatment [13, 14]. All the aforementioned fabrication methods and techniques are either expensive or can only be used for fabricating small sized samples. Triboelectric nanogenerators hold huge potential for commercial applications in the area of energy harvesting and sensing devices. But scalability of the fabrication techniques still remains a critical issue for mass production of triboelectric mechanism based devices. Moreover, scalability of fabrication processes is also important in order to realize large scale energy harvesters which can harvest mechanical energy from vehicles, human walking and ocean waves.

In case of mechanical energy harvesters which utilize piezoelectric and electromagnetic mechanism, the available fabrication techniques are not suitable for large scale manufacturing of devices due to high complexity and scalability problems. Mechanical energy harvesting devices based on triboelectric mechanism can potentially be manufactured in a scalable manner due to wide choice of materials which can be used to realize triboelectric mechanism based devices. In this chapter, we demonstrate a process flow for fabrication of triboelectric nanogenerators by fabricating components using large scale processing with high throughput. These components can then be integrated together to realize large scale triboelectric nanogenerator (LS-TENG). These LS-TENGs can be used for harvesting mechanical energy at a larger scale e.g. harvesting energy from vehicle motion by embedding on the roads, inside homes on the floor to harvest energy from human motion etc. Furthermore, roll-to-roll process based fabrication has also been proposed and demonstrated for fabricating large size pressure (force per unit

L. Dhakar, *Triboelectric Devices for Power Generation and Self-Powered Sensing Applications*, Springer Theses,
DOI 10.1007/978-981-10-3815-0_5

area) sensor arrays. Triboelectric mechanism based sensor arrays have earlier been demonstrated for self-powered tracking systems [15], tactile imaging [16] and displacement sensor systems [17]. These sensor arrays were fabricated by assembly of individual triboelectric devices and/or use of wafer level fabrication process for patterning the films. Assembly of individual triboelectric devices is not suitable for scaling up of manufacturing process to fabricate large size sensor arrays. On the other hand wafer level patterning of films can only be used for fabrication of small sized arrays. Here, we demonstrate a sensor array using large scale polymer films fabricated using roll-to-roll ultraviolet (UV) embossing process [18]. We also demonstrate a process flow for facile integration of these large scale polymer films into pressure sensor arrays. Experiments were conducted to demonstrate practical applications of these sensor arrays for self-powered motion tracking and posture monitoring. This work demonstrates scalable fabrication of TENGs and self-powered pressure sensor arrays, which will lead to extremely low cost and bringing them closer to commercial production.

5.2 Large Scale Energy Harvesting Using Roll-to-Roll Fabrication Process

5.2.1 Fabrication Process

For LS-TENG, patterned polyethylene terephthalate (PET) film and copper film are chosen as two triboelectric layers due to their relative tendency to attract and donate electrons, respectively. Roll-to-roll UV embossing [18] was used for the fabrication of patterned PET triboelectric layers. The schematic for setup for roll-to-roll fabrication process is shown in Fig. 5.1. The UV embossing system consists of four modules that are: (1) unwinding unit for supplying substrate film that is a blank PET film; (2) coating module for depositing UV curable resin on PET film; (3) UV embossing module for patterning microstructures on PET film and (4) rewinding module that provides a web tension for separating the embossed PET film from the embossing roller and then rewinds for collecting the embossed PET film. UV curable resin was coated on PET films using a slot die in the coating module and the coating thickness of 20–60 μm with thickness uniformity better than ±10% could be obtained. Large area patterned PET film is fabricated via roll-to-roll UV embossing inside UV embossing module with UV exposure. Figure 5.2a, b show images of the actual fabrication setup.

5.2.1.1 Fabrication of Mold

Large area flexible molds for roll-to-roll UV embossing were fabricated via patterning a thin photosensitive polyurethane sheet by means of designing micro patterns, photolithography and development of the photosensitive polyurethane

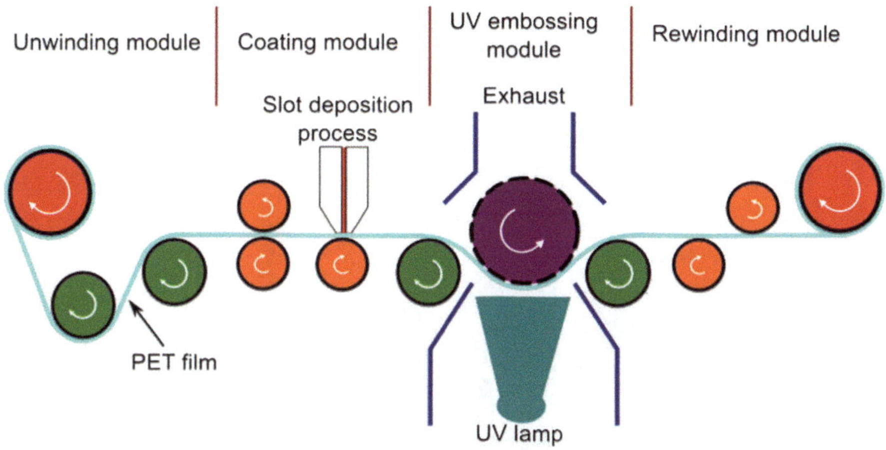

Fig. 5.1 Schematic of the fabrication setup for roll-to-roll UV embossing of large size polymer films

sheet in a chemical solution. The embossing roller was 500 mm long in axial direction with a diameter of 160 mm. The 1 mm thick patterned polyurethane sheet was then attached on the embossing roller by using double-sided adhesive tape. In order to obtain a roller mold with a seam as narrow as possible, the length in circumferential of the polyurethane flexible mold needs to cut precisely, taking into account of the thickness of double-side adhesive tape and thickness of the mold itself. A photograph of the fabricated embossing roller is shown in Fig. 5.3.

5.2.1.2 Roll-to-Roll UV Embossing of Patterned PET Films

PET film was pre-coated with a primer layer. The primer consisted of functionalized α-olefin-containing copolymers and cross-linking agents to improve the adhesion of coated UV sensitive resin layer during roll-to-roll UV embossing process. PET film with a thickness of 125 μm was first slot die-coated with a UV curable resin. The UV curable resin is a solvent-free acrylate-based lacquer with viscosity of about 100 *centipoise* at 25 °C. Thereafter the film was embossed using the embossing roller fabricated as described in the previous section. The UV lamp was irradiated directly from below the embossing-roller to cure the UV curable resin after the resin flows and fully fills the cavities on the mold. After the completion of coating, embossing and separation processes, micro patterns on the mold are transferred onto PET film. Hence, patterned PET film with micro patterns were obtained by means of roll-to-roll UV embossing. An optical image of a patterned PET film with line patterns (line pitch: 500 μm) is shown in Fig. 5.4a. A photograph of fabricated patterned PET film sample is shown in Fig. 5.4b.

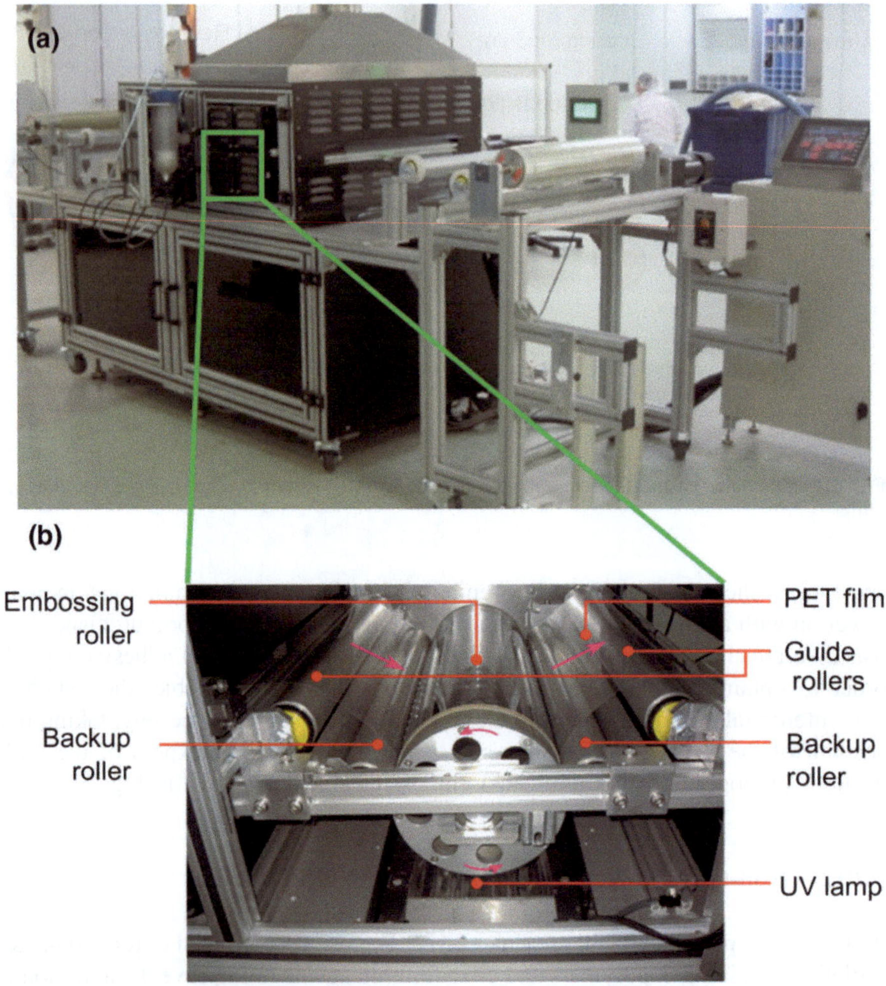

Fig. 5.2 **a** Photograph of the fabrication setup used for roll-to-roll UV embossing. **b** Labeled photograph of the UV embossing module

5.2.1.3 Assembly of LS-TENG

The patterned PET films using the roll-to-roll UV embossing act as one of the triboelectric layer in the large scale LS-TENG. A schematic of the assembled LS-TENG is shown in Fig. 5.4c. Patterned PET sheets were used to increase the contact area between the triboelectric layers. For the second triboelectric layer, an 18 μm thick copper film was attached on top of 50 μm thick liquid crystal polymer (LCP) substrate. This film was fabricated by laminating the copper film on LCP film. Thereafter 1 mm thick foam tape was used as spacer and assembled on top of patterned PET film. The copper film was then assembled on top of the spacers as

Fig. 5.3 A mold made of polyurethane-based photopolymer attached on embossing roller

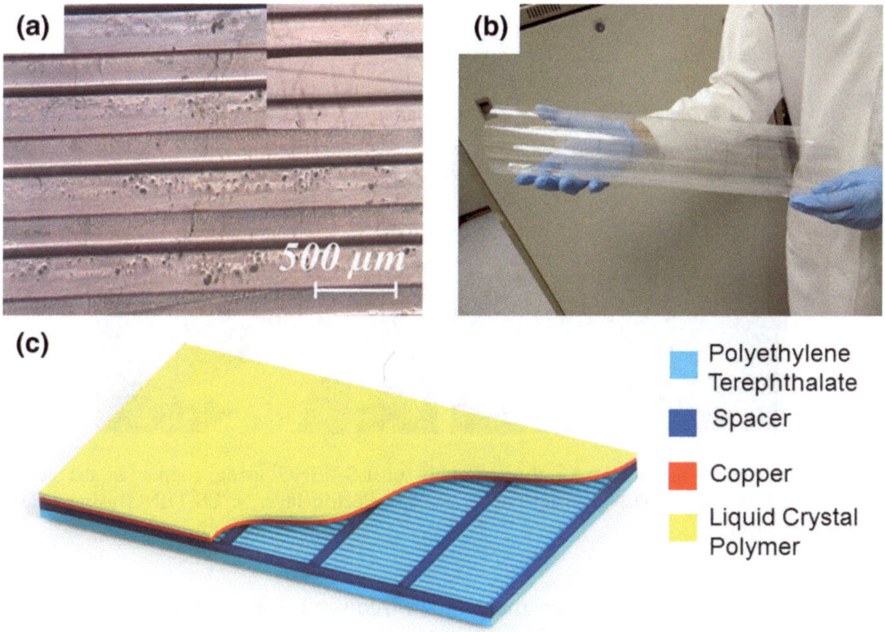

Fig. 5.4 **a** Optical image of the line patterns of fabricated on PET film using roll-to-roll UV embossing. **b** Photograph of a large scale patterned PET sheet fabricated using roll-to-roll UV embossing. **c** Schematic of LS-TENG

shown in Fig. 5.5. Copper film served dual purpose of a triboelectric layer and electrode to collect charges using an external load. Due to the large size of PET film, the film tends to sag down permanently onto the copper electrode which

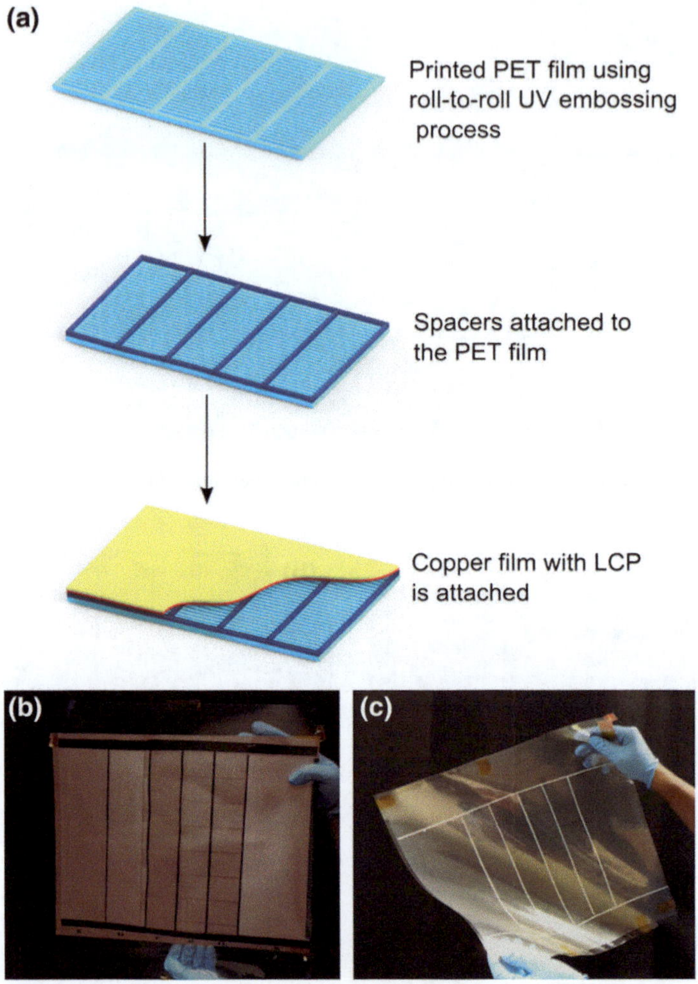

Fig. 5.5 **a** Detailed schematic for the assembly of LS-TENG using copper as the second triboelectric layer. **b** LS-TENG using copper as triboelectric layer. **c** LS-TENG using ITO as triboelectric layer

hinders the contact separation mechanism required for charge generation for mechanical energy harvesting. Therefore it is important to have spacers at regular spacing between copper and patterned PET triboelectric layer to help the large patterned polymer film rebound back to its original position as the device is subjected to mechanical force and released.

The same process was also used to fabricate transparent roll-to-roll based LS-TENG. To fabricate transparent LS-TENG, commercially available PET coated with indium tin oxide (ITO) film was also used and tested as second triboelectric layer. Figure 5.5a, b shows the photograph of the actual fabricated devices using copper and ITO as triboelectric layers respectively.

5.2.2 *Working Mechanism*

The schematic for operating mechanism of LS-TENG is shown in Fig. 5.6. Initially, both the triboelectric layers, i.e. patterned PET film and copper film are uncharged as depicted in the initial state in Fig. 5.6. During the device operation, the copper film was kept at bottom while patterned PET film is at the top separated by spacers from copper film. As there is any mechanical force applied on the PET film due to activities like walking, palm tapping etc., the electrons are injected from the copper film into PET layer owing to copper's higher tendency to donate electrons relative to PET. The charges on the PET film are preserved due to its insulating properties, whereas the charges on conductive copper film can flow through the load resistor connected to the ground as electric potential changes at the copper electrode. After the applied force is released, the negative charges embedded on the PET film start separating from the copper film. This results into an increase in the potential at the copper electrode, leading to the flow of electrons from the reference ground electrode towards the copper electrode. As the PET film separates due to the stored elastic energy stored when applied with a mechanical force, it reaches a maximum point of separation between the two triboelectric layers. At the maximum separation point, there is no flow of electrons in the load resistor connected between electrode and ground. Thereafter, the PET film starts approaching towards the copper film leading to the flow of electrons in the load resistor in opposite direction. This is equivalent of the one cycle of electricity generation due to application of force on LS-TENG.

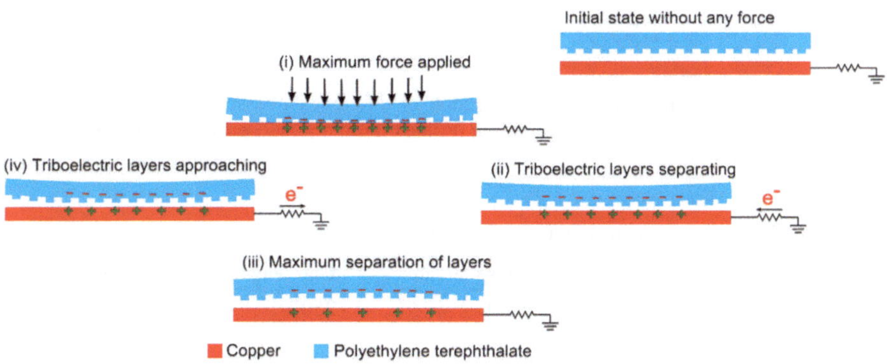

Fig. 5.6 Working mechanism of LS-TENG

5.2.3 Effect of Different Embossed Patterns

To study the effect of embossed patterns on the performance of triboelectric mechanism, different samples were fabricated. We fabricated two types of line patterns and three types of square shaped patterns. The topography of these five patterns are shown in Fig. 5.7a–e. The geometric parameters for the line and square patterns are given in Table 5.1. As shown in the cross section profiles in Fig. 5.7a–e, side walls of embossed structures are not perfectly vertical due to fabrication process limitation. Also, the top of the embossed structure was observed to be curved instead of completely flat. This results in reduction of effective area at the top of embossed structures as compared to the base area of the embossed structures. At the same time, embossing of structures is important as microstructured films results into improved performance of the triboelectric performance. This is due to easier charge separation and enhanced triboelectric effect exhibited by the

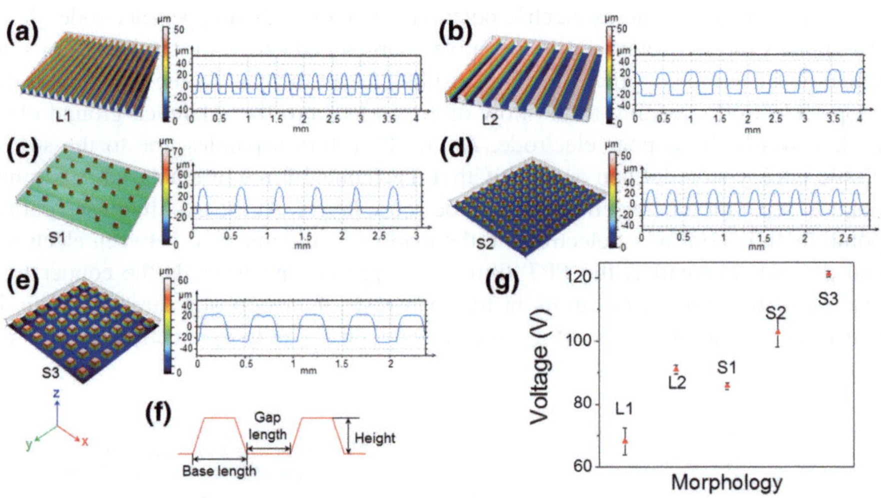

Fig. 5.7 **a–e** 3-D morphology and cross section of different embossed patterns. **f** Geometrical parameters of different embossed patterns. **g** Voltage output generated by the different patterns under same force level

Table 5.1 Geometrical parameters for different fabricated morphologies

	Pattern	Base length	Gap length	Height (μm)
Line patterns	L1	250	250	~45
	L2	500	500	~45
Square patterns	S1	300	700	~45
	S2	300	200	~45
	S3	600	400	~45

patterned films as demonstrated by Fan et al. These two effects work in contrast with each other affecting the performance of TENG. The samples were tested using a mechanical stage to examine the performance of different morphologies. The order of the performance of different morphologies was observed as under:

$$L1 < S1 < L2 < S2 < S3$$

Amongst L1 and L2, the designed base area of the structures was same but because of tapered shape of fabricated structures due to process limitation, the effective contact area at the top was reduced for L1. This resulted in better performance of L2 compared to L1 in spite of same base area of embossed structures in both the cases. Amongst S1, S2 and S3, S1 was designed to have minimum base area and was observed to have the least performance. S2 and S3 were designed to have same base area but due to fabrication process limitation, the S2 had a lesser top contact area as compared to S3. It was observed that S3 had a significantly better performance compared to S2 which is in accordance with the results observed for L1 and L2.

5.2.4 Applications in Energy Harvesting

The energy harvesting characteristics were measured using palm tapping on LS-TENG. The peak-to-peak output voltage was measured to be 350 V when tapped using hand (Fig. 5.8a) using copper and patterned PET film as a triboelectric pair. The peak-to-peak short circuit current generated by LS-TENG was measured to be 20 μA as shown in Fig. 5.8b. To study the variation of power transferred to a load resistor, LS-TENG was connected to different load resistors. The peak voltage measured across the load resistor connected across LS-TENG increases continuously as the value of load resistance increases as shown in Fig. 5.8c. On the other hand, the current flowing across the resistor decreases as the value of load resistor decreases. The peak power density using the active area used for energy generation, was measured to be 62.5 mW m^{-2}. Using ITO and patterned PET as a triboelectric pair, the peak-to-peak output voltage and current generated were measured to be 310 V and 19.6 μA. The peak power density from the device was measured to be 53.82 mW m^{-2}. The optimum performance of Cu based LS-TENG was observed to be slightly better than ITO based LS-TENG device.

The voltage and current output of LS-TENG were also tested at different tapping frequencies from 1 to 5 Hz. Both voltage and current output were observed to increase as the frequency of applied force was increased as shown in Fig. 5.9a, b. The peak-to-peak voltage increased from 232 to 486 V as the frequency increased from 1 to 5 Hz. The peak-to-peak current increased from 7.2 to 21.2 μA as the frequency increased from 1 to 5 Hz. The increase in voltage output with increased

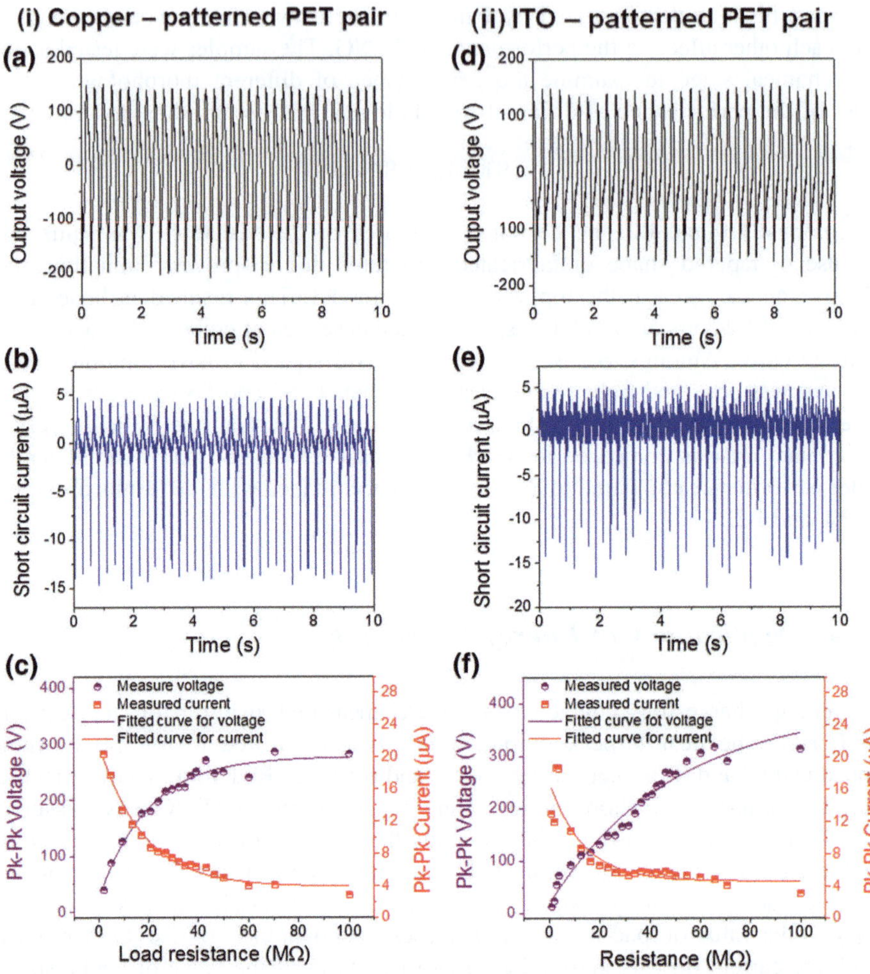

Fig. 5.8 **a** Output voltage and **b** current generated by using LS-TENG using copper as second triboelectric layer. **c** Dependence of peak-to-peak (Pk-Pk) voltage and current on load resistance. **d** Output voltage and **e** current generated by using LS-TENG using ITO as second triboelectric layer. **f** Dependence of peak-to-peak (Pk-Pk) voltage and current on load resistance

frequency can be explained due to increased value of applied force as the tapping frequency increased. The higher applied force led to increased contact area between the triboelectric layers resulting in higher voltage output. The increase in the current output can be explained by the combined effect of the increased force and the reduction of the cycle time for the charges to flow in the external circuit as the tapping frequency increased.

To demonstrate the viability of LS-TENG in day-to-day applications, the device was assembled at the back of an office chair as shown in Fig. 5.10a. The output was

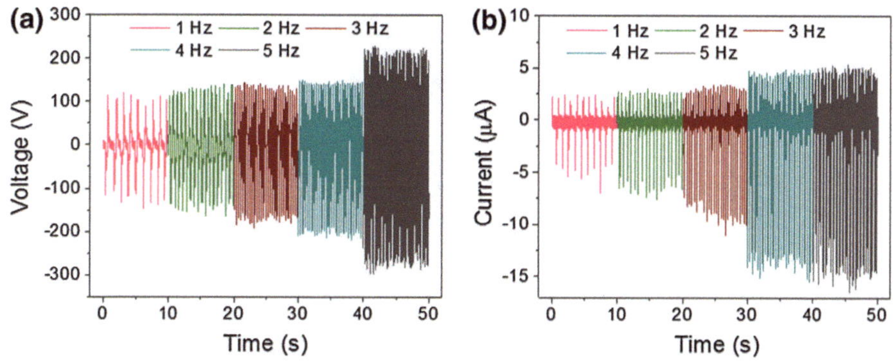

Fig. 5.9 **a** Voltage generated by LS-TENG at different frequencies. **b** Current generated by LS-TENG at different frequencies

recorded as the subject sat on chair in three conditions: high activity, low activity and no activity. A commercial 3-axis accelerometer ADXL325 from Analog Devices, Inc. was assembled at the back of the chair to observe the activity levels. The accelerometer data for 3 axes is shown in Fig. 5.9b, c along with output voltage and current data. LS-TENG generated peak-to-peak voltage and current of up to 85 V and 0.75 μA, respectively in the high activity region. During low activity, LS-TENG produced peak-to-peak voltage and current of up to 35 V and 0.3 μA, respectively. LS-TENG was also used to light up nearly 90 commercial LEDs by directly connecting them to LS-TENG subjected to hand tapping (Fig. 5.10d).

5.3 Triboelectric Pressure Sensor Arrays Using Roll-to-Roll UV Embossed Films

5.3.1 Fabrication and Assembly

For the sensor array, square patterns were fabricated using roll-to-roll UV embossing on the PET film in a 7 × 3 array using roll-to-roll process. We used cover stock paper as a substrate to pattern electrodes of 2.5 cm × 3 cm using aluminum foil. Both the components were used to keep the cost low for the sensor array. PET sheet with 200 μm were used as spacers between the patterned PET film and aluminum electrodes to maintain the gap between the triboelectric layers. Spacers were also used to separate all the pixels from each other to minimize the cross-talking between pixels when the sensor array is subjected to external force/pressure. Figure 5.11 shows a schematic diagram of the force/pressure sensor array.

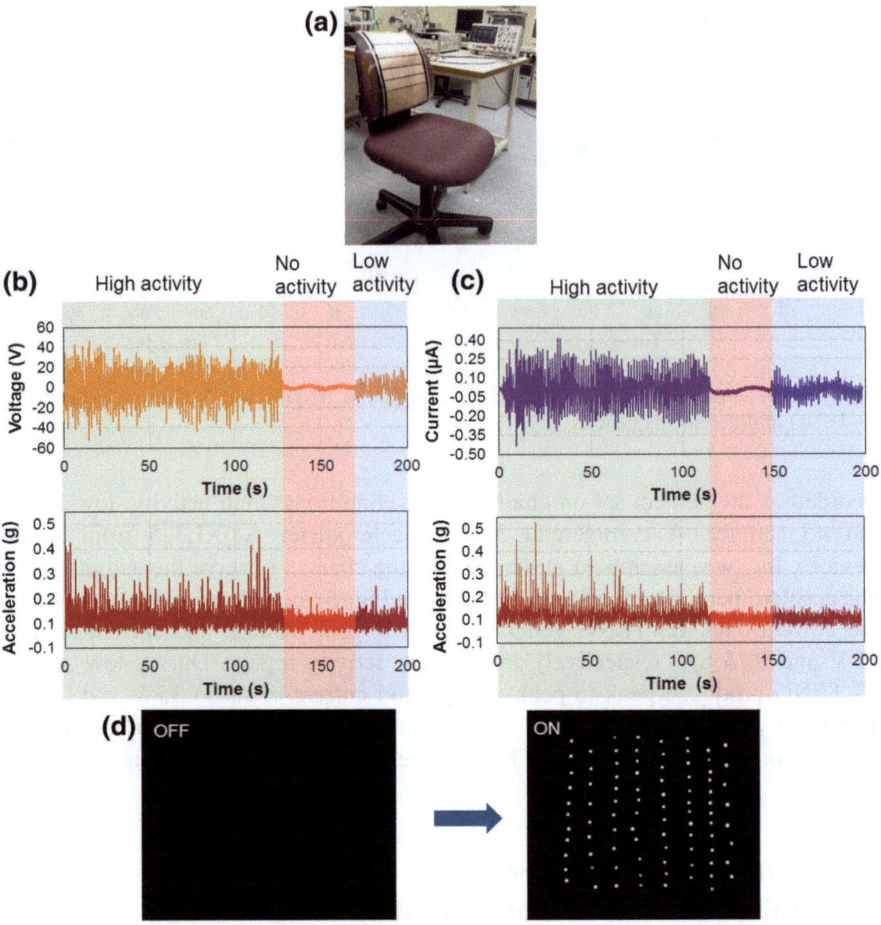

Fig. 5.10 **a** LS-TENG assembled on the back of a chair. **b** Voltage and **c** Current signal generated by LS-TENG by natural movement of a human working on computer. The data below the voltage and current data were measured using an ADXL325 assembled at the back of chair. **d** Photographs of LEDs in OFF state and ON state after being lighted up by mechanical energy harvested using hand tapping

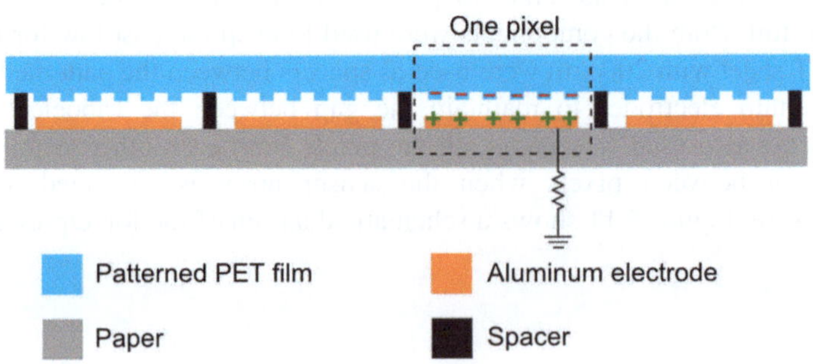

Fig. 5.11 Schematic representation of the pressure sensor array

5.3.2 Sensor Array Characterization

The fabricated sensor array can be used for motion tracking, tactile sensing, posture monitoring and various other applications. A schematic of 7 × 3 sensor array is shown in Fig. 5.12a. A photograph of the fabricated sensor array is shown in Fig. 5.12b. For the sensor array measurement, for every pixel, a load resistor of

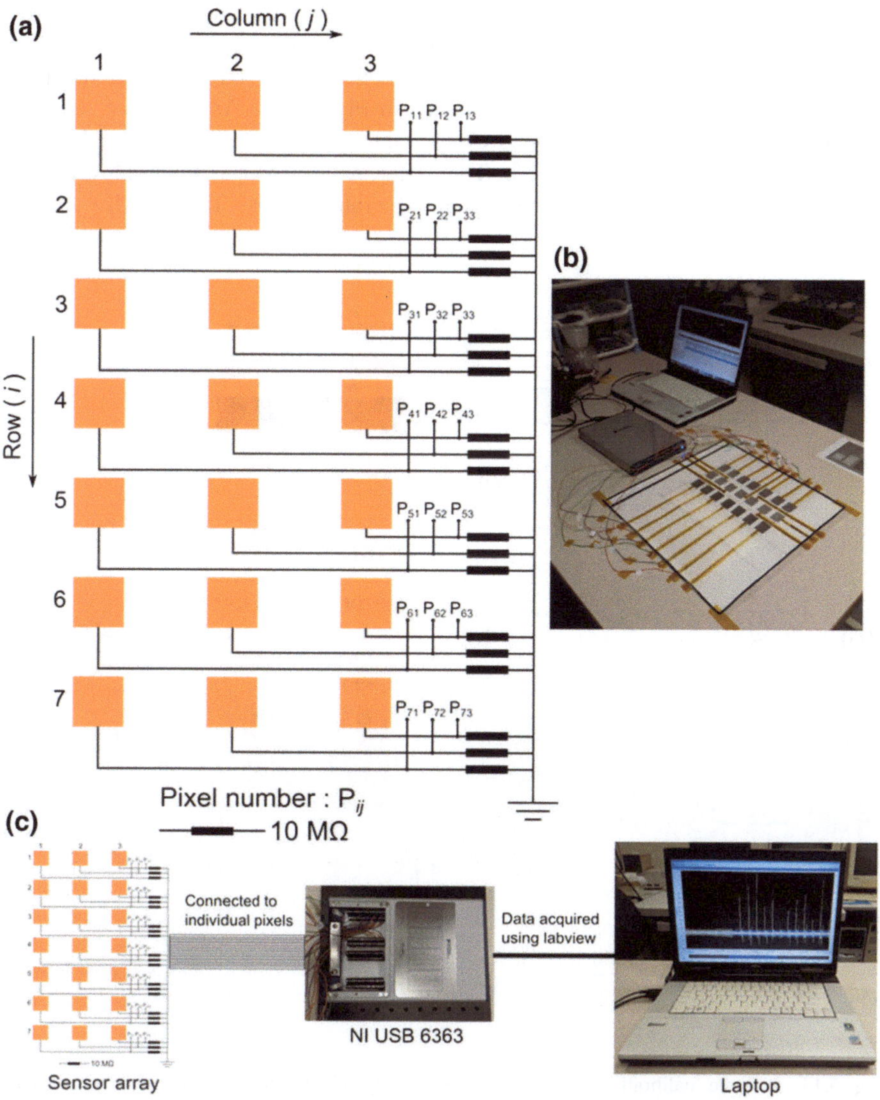

Fig. 5.12 **a** Schematic diagram of the triboelectric pressure sensor array. **b** Photograph of the fabricated large size pressure sensor array. **c** Data acquisition setup for the sensor array

10 MΩ is connected between the aluminum electrode and common ground. The output across these load resistors were measured using individual channels of USB-6363 data acquisition (DAQ) board from National Instruments. The output from 21 channels from USB-6363 was acquired using Lab View to observe the individual output from every pixel (as shown in Fig. 5.12c).

The pixels were tested using a commercial force sensor for calibration. The schematic for setup is shown in Fig. 5.13a. As the pixel was stimulated using difference levels of force, the signals were collected from both pixel output and the force sensor. Time domain signals generated by the pixel and force sensor were plotted on same time scale in Fig. 5.13b as the force was applied on the pixel. As the pixel is pressed, initially there is a negative peak observed in the voltage signal generated across the load resistor as shown in Fig. 5.13b. As the force reaches a maximum value, the two triboelectric layers are in contact and measured voltage across 10 MΩ resistor reaches zero. Thereafter, as the force is released, a positive peak is observed in the voltage signal generated across the 10 MΩ resistor.

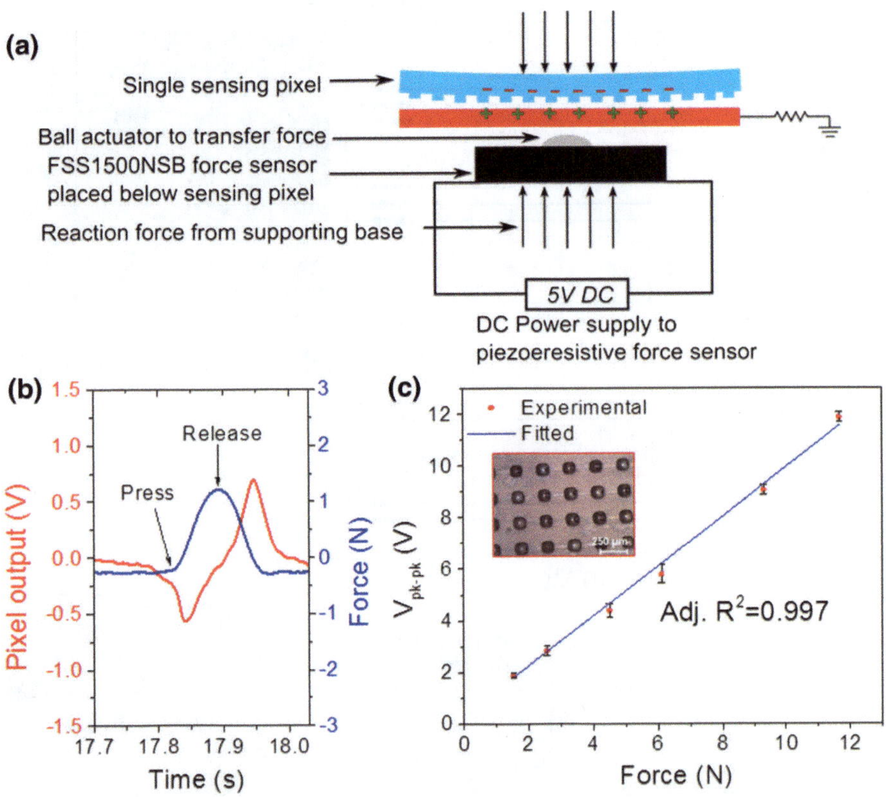

Fig. 5.13 **a** Force calibration setup for sensor array pixel using commercial force sensor FSS1500NSB from Honeywell. **b** Time domain signal generated by the pixel and the force sensor beneath the pixel. **c** Peak-to-peak voltage generated by the pixel for different values of applied forces. *Inset* image shows optical image of the micropillar structures fabricated on PET sheet

The peak-to-peak voltage generated by the pixel are plotted for different values of force applied as shown in Fig. 5.13c. The peak-to-peak voltage was observed to increase in a linear manner as the force level applied to the pixel increases. The force sensitivity of the sensing pixel was characterized to be 0.935 V N^{-1} which was equivalent to pressure sensitivity of 1.33 V kPa^{-1}.

5.3.3 Motion Tracking and Security Applications

The sensor array was tested for motion tracking of a cylindrical object rolling over the array. A cylindrical object with a diameter of 9.2 cm, width of 1.9 cm and weight of 30 g, was rolled on a linear array of six pixels from P1 to P6 as shown in Fig. 5.14. For each pixel, the time domain output voltage signals are plotted.

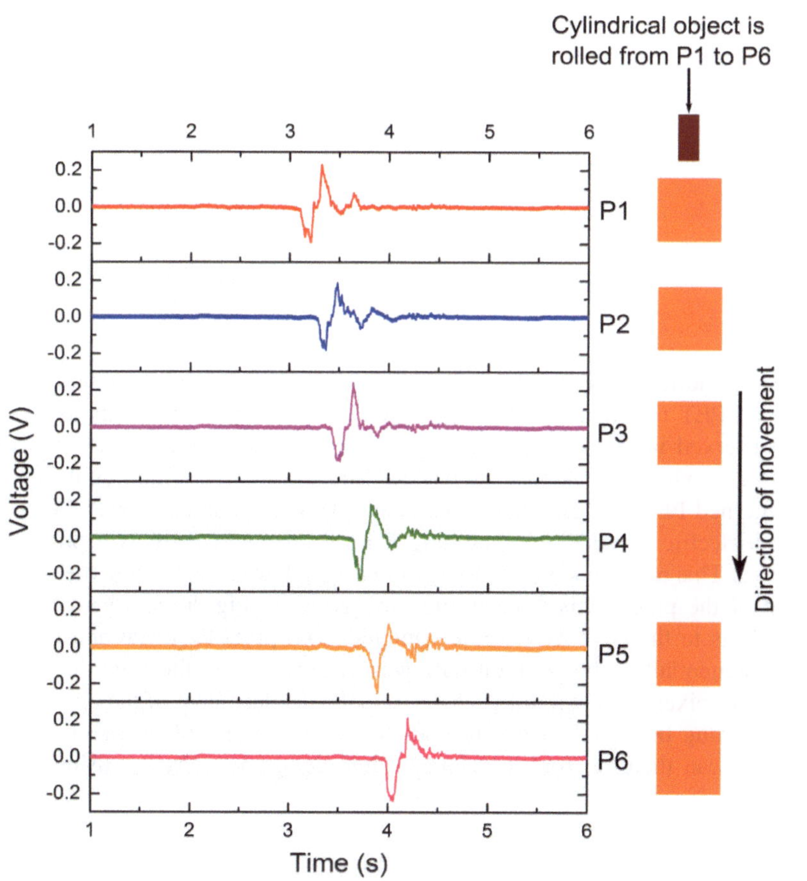

Fig. 5.14 Time domain output voltage signals generated by an array of 6 pixels as a cylindrical object is rolled over them. The pixels generate a signal as the object contacts the pixel

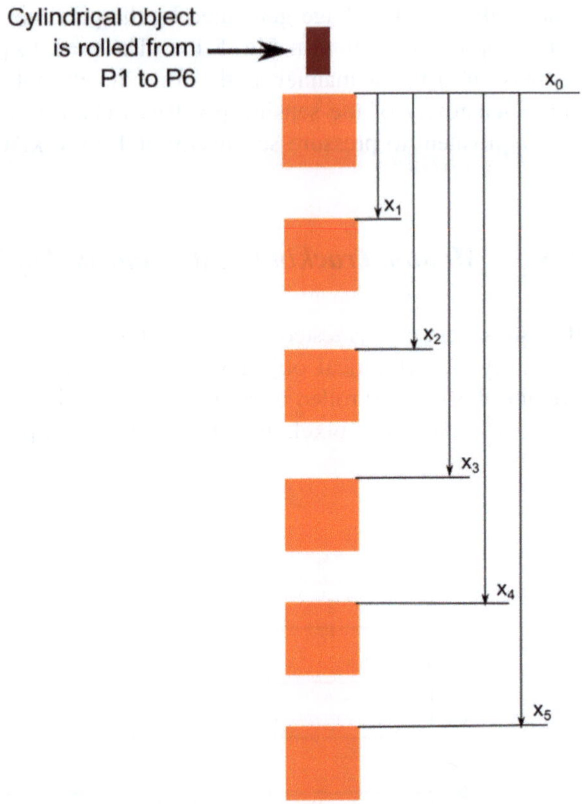

Fig. 5.15 Calculation of velocity and acceleration data as the sensor array is used as a self-powered movement tracking system

As the cylindrical object rolls over the sensors array, it leads to contact between patterned PET film and the aluminum electrodes. This in turn leads to a negative peak observed in the voltage signal generated by the pixel. As the object keeps rolling, a maximum peak force is reached and there is maximum contact between the patterned PET film and aluminum electrode. As the object leaves the pixel and the triboelectric layers start separating from each other, there is a positive peak observed. This mechanism leads to a voltage signal generated, when the object rolls on top of the pixel. This is clear from the graph in Fig. 5.13, where the voltage pulse shifts to the right as the object moved from pixel P1 to P6 and touches the pixels sequentially. The position data was acquired using the time domain signal from every pixel as shown in Fig. 5.13. x_i is the displacement of the object from the initial starting position when it reaches the starting point of ith pixel (Fig. 5.15). This data can then be used to calculate the average velocity v_{av} using following equation:

$$v_{av,i} = \frac{x_{i+1} - x_i}{\Delta t_i} \quad (5.1)$$

Table 5.2 Calculated values of average velocity and average acceleration

	Distance (cm)	Time (s)	Average velocity (cm s^{-1})	Average acceleration (cm s^{-2})
x_0	0	3.10	0	–
x_1	3.7	3.29	19.68	0.01
x_2	7.4	3.43	26.43	0.02
x_3	11.1	3.65	16.82	−0.02
x_4	14.8	3.80	24.50	0.02
x_5	18.5	3.98	20.11	−0.04

where, $v_{av,i}$ is the average velocity of the object between P_{i-1} and P_i; and Δt_i is the time interval elapsed between P_{i-1} and P_i.

Similarly average acceleration $a_{av,i}$ can be calculated using the average velocity values calculated using Eq. (5.2):

$$a_{av,i} = \frac{v_{i+1} - v_i}{\Delta t_i} \tag{5.2}$$

where, $a_{av,i}$ is the average velocity of the object between P_{i-1} and P_i; and Δt_i is the time interval elapsed between P_{i-1} and P_i. Equations (5.1) and (5.2) were used to calculate the average velocity and acceleration values as shown in Table 5.2.

The sensor array pixels were also demonstrated to respond to very small forces e.g. using the air blown from human mouth as shown in Fig. 5.16. These low cost and large scale sensor arrays can potentially be used for movement tracking of objects with applications in security systems, sports/athletic training and remote patient monitoring.

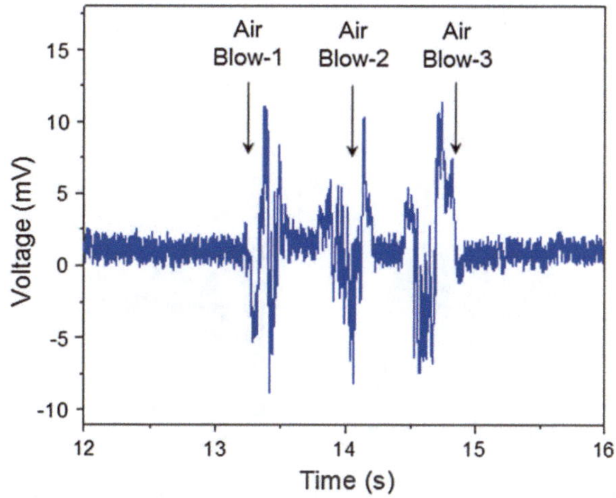

Fig. 5.16 Output generated using blowing of air using human mouth on a single pixel

5.3.4 Applications in Posture Tracking

The sensor array was also demonstrated for application in posture monitoring. For experiments, the array was assembled at the back of a chair as shown in Fig. 5.17g. The sensor array pixels generated an electrical output as the subject touches the

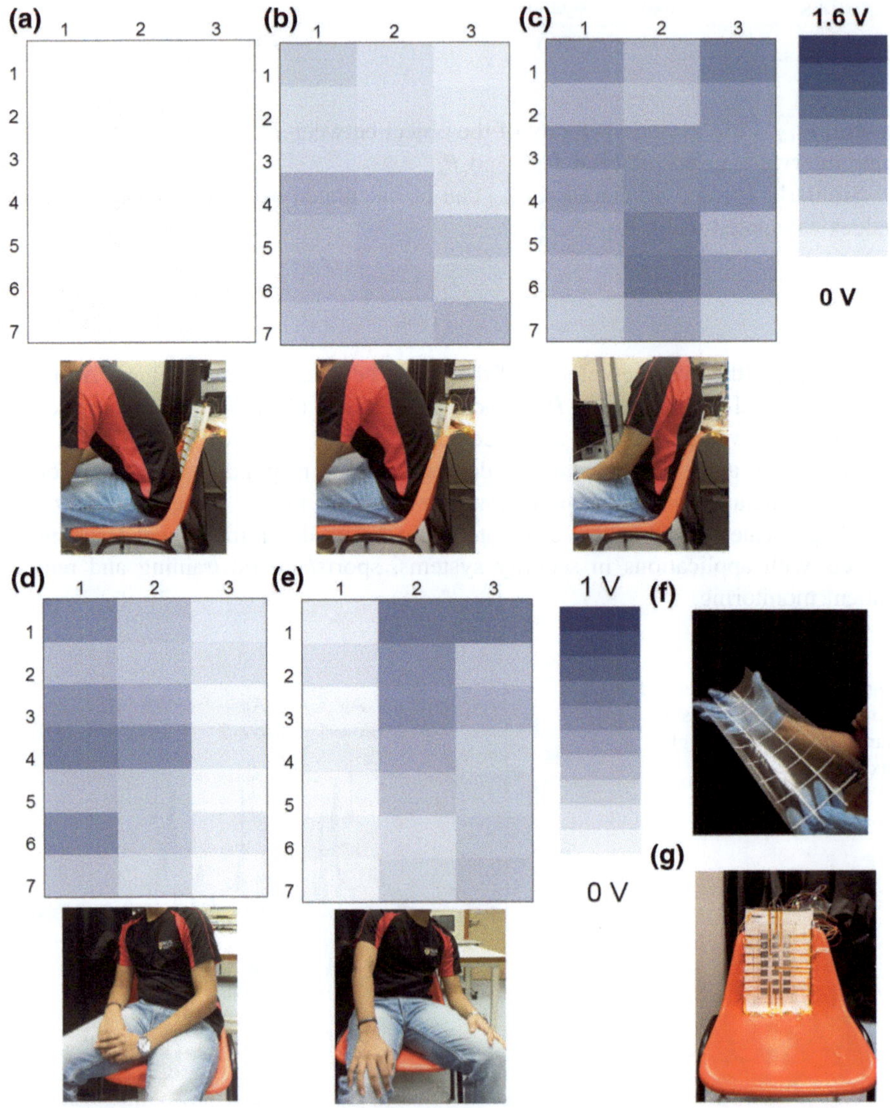

Fig. 5.17 Heat map of a 7 × 3 sensor array for different posture when array assembled on back of chair. **a** Without touching the sensor array **b** partially touching the sensor array pixels **c** completely touching the sensor array pixels **d** leaning on right side of the subject **e** leaning on left side of the subject. **f** Photograph of PET film fabricated using roll-to-roll fabrication process assembled with spacers. **g** Sensor array assembled on the back of a chair

pixels of sensor array. The heat map of the peak-to-peak voltage generated by the pixels is shown in Fig. 5.7a–e for different positions of the subject. In position (a) the subject is not touching any pixels; as the subject moves to position (b), the subject contacts the majority of pixels in the lower half of sensor array (after row 3) that results in darker pixels in lower half as compared to the upper half. Also, it can be observed that there is some output observed in the pixels above row 3 as compared to the position (a). This can be attributed to increase in the overall noise level and cross-talking between pixels, as the PET film deforms when applied with a external force. In position (c), the subject straightens the back and pressure is applied on whole array. Thereafter, Fig. 5.16d, e show the results as the subject tilts to right and left respectively (as seen from subject's position). The experiments demonstrate that the sensor array can be used for applications like posture monitoring, sleep pattern monitoring, walking pattern monitoring and electronic skin applications.

5.4 Summary

We have developed and demonstrated a process flow for large scale fabrication triboelectric nanogenerators with high throughput. Roll-to-roll UV embossing has been utilized to fabricate large size patterned polymer films to realize triboelectric nanogenerators. LS-TENG generated a peak-to-peak voltage and current of 350 V and 20 μA respectively. LS-TENG was demonstrated to generate a power of 62.5 mW m^{-2} using palm tapping. Roll-to-roll fabrication based process was also used to fabricate large size and low cost pressure sensor array with a detection sensitivity of 1.33 V kPa^{-1}. We also conducted experiment using the sensor array for posture monitoring by assembling the array at the back of chair. The fabricated large size sensor array can be used in motion tracking, patient monitoring, posture monitoring and electronic skin applications. This work is aimed at developing and demonstrating scalable fabrication process for large scale production of triboelectric nanogenerators and sensor arrays with high throughput and low cost.

References

1. F.-R. Fan, L. Lin, G. Zhu, W. Wu, R. Zhang, Z.L. Wang, Transparent triboelectric nanogenerators and self-powered pressure sensors based on micropatterned plastic films. Nano Lett. **12**, 3109–3114 (2012)
2. L. Dhakar, F. Tay, C. Lee, Investigation of contact electrification based broadband energy harvesting mechanism using elastic PDMS microstructures. J. Micromech. Microeng. **24**, 104002 (2014)
3. L. Dhakar, F.E.H. Tay, L. Chengkuo, Development of a broadband triboelectric energy harvester with SU-8 micropillars. J. Microelectromech. Syst. **24**, 91–99 (2015)

4. X.-S. Zhang, M.-D. Han, R.-X. Wang, F.-Y. Zhu, Z.-H. Li, W. Wang et al., Frequency-multiplication high-output triboelectric nanogenerator for sustainably powering biomedical microsystems. Nano Lett. **13**, 1168–1172 (2013)
5. B. Meng, W. Tang, Z.-H. Too, X. Zhang, M. Han, W. Liu et al., A transparent single-friction-surface triboelectric generator and self-powered touch sensor. Energy Environ. Sci. **6**, 3235–3240 (2013)
6. S. Wang, Y. Xie, S. Niu, L. Lin, Z.L. Wang, Freestanding triboelectric-layer-based nanogenerators for harvesting energy from a moving object or human motion in contact and non-contact modes. Adv. Mater. **26**, 2818–2824 (2014)
7. L. Zhang, L. Cheng, S. Bai, C. Su, X. Chen, Y. Qin, Controllable fabrication of ultrafine oblique organic nanowire arrays and their application in energy harvesting. Nanoscale **7**, 1285–1289 (2015)
8. J. Yang, J. Chen, Y. Liu, W. Yang, Y. Su, Z.L. Wang, Triboelectrification-based organic film nanogenerator for acoustic energy harvesting and self-powered active acoustic sensing. ACS Nano **8**, 2649–2657 (2014)
9. G. Zhu, Y. Su, P. Bai, J. Chen, Q. Jing, W. Yang et al., Harvesting water wave energy by asymmetric screening of electrostatic charges on a nanostructured hydrophobic thin-film surface. ACS Nano **8**, 6031–6037 (2014)
10. G. Zhu, Z.-H. Lin, Q. Jing, P. Bai, C. Pan, Y. Yang et al., Toward large-scale energy harvesting by a nanoparticle-enhanced triboelectric nanogenerator. Nano Lett. **13**, 847–853 (2013)
11. C.K. Jeong, K.M. Baek, S. Niu, T.W. Nam, Y.H. Hur, D.Y. Park et al., Topographically-designed triboelectric nanogenerator via block copolymer self-assembly. Nano Lett. **14**, 7031–7038 (2014)
12. D. Kim, S.-B. Jeon, J.Y. Kim, M.-L. Seol, S.O. Kim, Y.-K. Choi, High-performance nanopattern triboelectric generator by block copolymer lithography. Nano Energy **12**, 331–338 (2015)
13. Z.-H. Lin, Y. Xie, Y. Yang, S. Wang, G. Zhu, Z.L. Wang, Enhanced triboelectric nanogenerators and triboelectric nanosensor using chemically modified TiO_2 nanomaterials. ACS Nano **7**, 4554–4560 (2013)
14. S.-H. Shin, Y.H. Kwon, Y.-H. Kim, J.-Y. Jung, M.H. Lee, J. Nah, Triboelectric charging sequence induced by surface functionalization as a method to fabricate high performance triboelectric generators. ACS Nano **9**, 4621–4627 (2015)
15. Y. Yang, Y.S. Zhou, H. Zhang, Y. Liu, S. Lee, Z.L. Wang, A single-electrode based triboelectric nanogenerator as self-powered tracking system. Adv. Mater. **25**, 6594–6601 (2013)
16. L. Lin, Y. Xie, S. Wang, W. Wu, S. Niu, X. Wen et al., Triboelectric active sensor array for self-powered static and dynamic pressure detection and tactile imaging. ACS Nano **7**, 8266–8274 (2013)
17. Y. Yang, H. Zhang, J. Chen, Q. Jing, Y.S. Zhou, X. Wen et al., Single-electrode-based sliding triboelectric nanogenerator for self-powered displacement vector sensor system. ACS Nano **7**, 7342–7351 (2013)
18. Z. Zhong, X. Shan, Microstructure formation via roll-to-roll UV embossing using a flexible mould made from a laminated polymer–copper film. J. Micromech. Microeng. **22**, 085010 (2012)

Chapter 6
Triboelectric Mechanism for Bidirectional Tactile Sensing and Energy Harvesting

6.1 Motivation

Triboelectric effect has been utilized to realize sensors for various applications e.g. vibration sensing [1], chemical sensing [2–4] and tactile sensing [5–7]. Tactile sensors can be used to sense various type of physical interactions. Various tactile sensors have been proposed based on capacitance [8, 9] and resistance [10, 11], piezoelectric effect [12] and piezophotonic effect [13]. But majority of these sensors have an ability to sense mechanical stimulations in normal direction but unable to sense the shear force stimulations, which severely limits in their application in practical situations. Therefore it becomes imperative to design sensors which can be used for multi directional tactile sensing to gather richer information from the physical interactions. In many of the applications in robotics and electronic skin, it is essential to differentiate between the normal and sliding motion. In this chapter, we propose a triboelectric mechanism based design that can be utilized for both normal and shear tactile sensing. These devices can also be utilized for multidirectional mechanical energy harvesting.

6.2 Design and Fabrication

6.2.1 Device Configuration

To enable bidirectional sensing, the device should generate distinct electrical signals for in-plane and out-of-plane stimulations. For multidirectional tactile sensing using triboelectric effect, a device design is proposed and fabricated as shown in

L. Dhakar, *Triboelectric Devices for Power Generation and Self-Powered Sensing Applications*, Springer Theses,
DOI 10.1007/978-981-10-3815-0_6

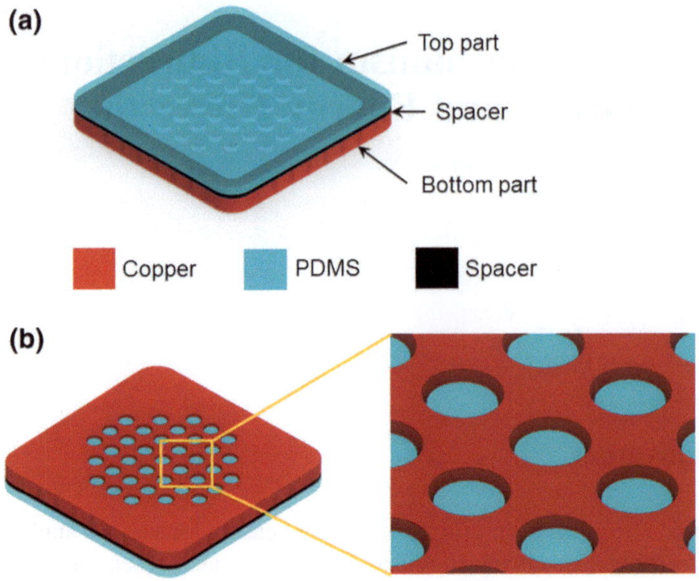

Fig. 6.1 **a** Schematic of the bidirectional triboelectric tactile sensing device. **b** Flipped view of the device. The *inset* shows the pillar inserted inside the holes

Fig. 6.1. The device comprises of two parts: top and bottom. The top part consists Si substrates etched with through holes. The bottom parts consists of PDMS micropillars which are assembled to inserted through holes in Si substrate.

The PDMS layer and the copper film coated Si substrate comes in contact with each other when mechanical force is applied on the device. The difference in direction of the applied force leads to different interaction between the top and bottom part. This difference in interaction leads to difference in electrical signals generated from the device. This mechanism is explained in next section in detail.

6.2.2 Working Principle

The device has been designed to generate response for both normal and shear force. The device works on the principle of generating surface charges using triboelectric effect, when the triboelectric layers are put in contact with each other due to external mechanical stimulation. The schematic for the working mechanism of the bidirectional tactile sensor is shown in Fig. 6.2. The surface area of the PDMS layer can be divided in two parts: flat surface area and surface area in the form of tapered pillars as shown in Fig. 6.2a. The inherent structure of the device results into

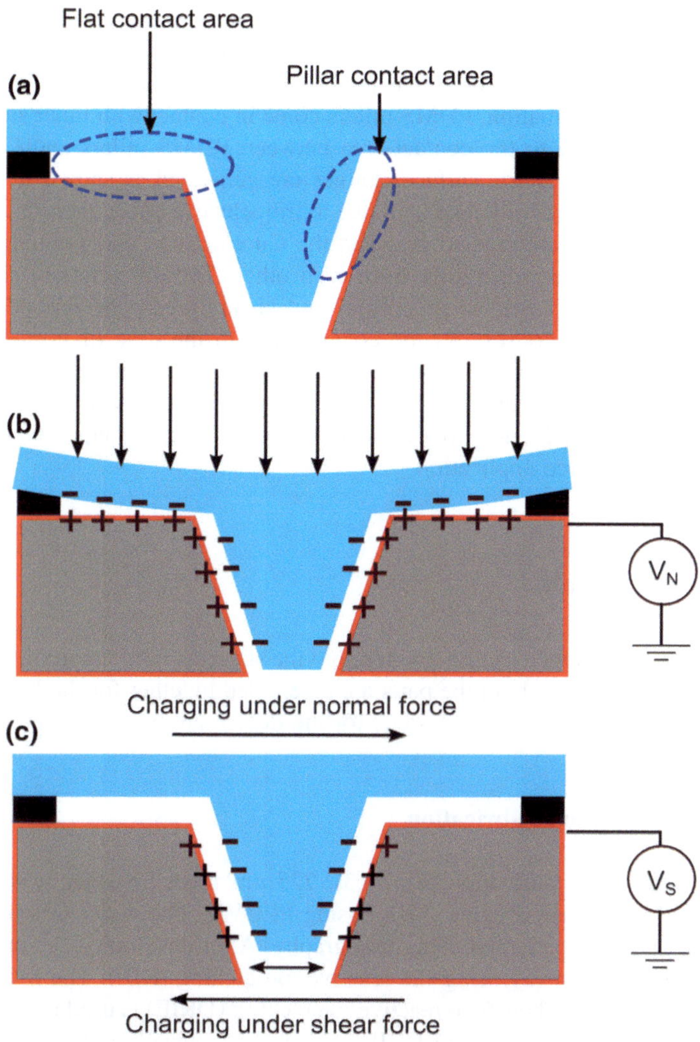

Fig. 6.2 **a** Two different types of contact areas labeled on the PDMS triboelectric layer. **b** Surface charges generated due to triboelectric effect due to out-of-plane/normal force. **c** Surface charges generated due to triboelectric effect due to in-plane/shear force

difference in interaction between the PDMS layer and Cu film, for different mechanical stimulation. This difference in interaction can be used to understand the working principle of the proposed device.

When normal force is applied on the top elastic PDMS film, it presses the film towards the Cu film in the normal direction. This results in the deformation of the

PDMS layer and it comes in contact with Cu film in both flat area and pillar area. This leads to generation of surface charge in the contacted regions as shown in Fig. 6.2b. In contrast, when the shear force is applied on the PDMS film, the film undergoes shear deformation. PDMS pillars come in contact with inner sidewalls of the etched holes. Due to the contact area between PDMS pillars and Cu film on inner sidewalls, triboelectric surface charges are generated as shown in Fig. 6.2c. As these surface charges are generated due to triboelectric effect, there is an electric potential V_N and V_S are created between the Cu electrode and ground when the triboelectric layers start separating from each other. The difference in charge distribution in both the cases when normal and shear forces are applied, leads to different electrical responses. As the contact area in the case of normal force is significantly higher compared to shear force application, the normal magnitude of V_N is expected to be higher than V_S. Also the variation of V_N would be different from V_S characteristics due to difference in mechanical characteristics during normal and shear force application.

6.2.3 *Fabrication*

The fabrication of device can be divided into the separate fabrication of the top and bottom parts. Thereafter both the parts are assembled together for the fabrication of complete device. The fabrication steps for the device are explained as follows.

6.2.3.1 Bottom Part Fabrication

For the bottom part fabrication (Fig. 6.3), 725 μm thick Si wafer was deposited with 2 μm thick silicon dioxide (SiO_2) using PECVD. The wafer was then coated with photoresist and patterned using lithography. After developing the patterns, the SiO_2 layer is etched away using the reactive ion etching (RIE). The SiO_2 patterns acted as a hard mask. Then deep reactive ion etching (DRIE) is used to etch away Si up to 400 μm. As the device are DRIE etched, the sidewall angle is a tilted at an angle of nearly 83°. Thereafter, the photoresist layer is stripped and SiO_2 is etched away. In the final step of creating through holes, the samples are background up to 300 μm. An optical image of the fabricated holes is shown in Fig. 6.3b.

Then the samples are coated with 10 nm/100 nm thick Cr/Cu layer on both the sides of the samples. Cr is coated to improve the adhesion between the Si substrate and the electrode. As the sidewalls are tilted, it results into uniform Cr/Cu layer on the inside of the holes. This was verified by the inspection of the cross section of the holes under optical microscope as shown in Fig. 6.4. As the sidewalls are coated with Cr/Cu layer, both the top side and bottom side of the Si substrate are

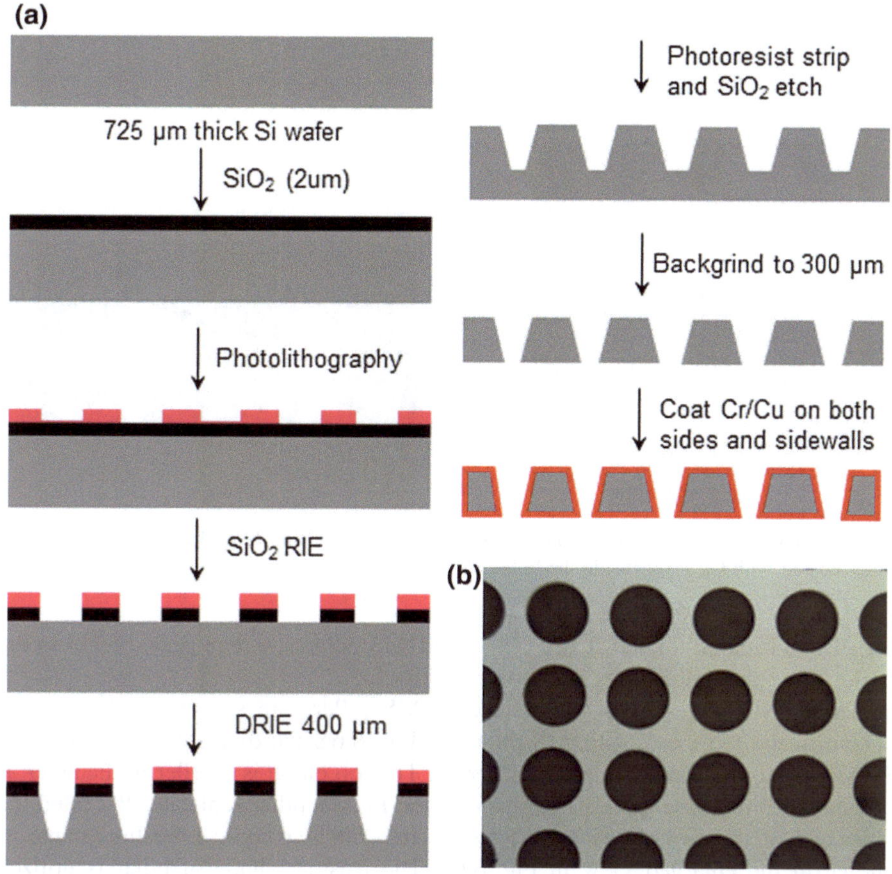

Fig. 6.3 **a** Process flow for the fabrication of the bottom part. **b** Optical image of through holes created in Si substrate

electrically connected. The Cu layer on Si serves as a triboelectric layer to generate surface charges and also an electrode to power the external load connected to TENG.

6.2.3.2 Top Part Fabrication

The top part comprises of the PDMS micropillars. For PDMS micropillar fabrication, the Si samples with etched through holes are attached to a glass slide. These Si chips attached to a glass slide were used as a mold to fabricate high aspect ratio PDMS micropillar structures. 10:1 PDMS mixture was poured on the sample and placed in a vacuum chamber to degas and let the PDMS enter the holes. Thereafter,

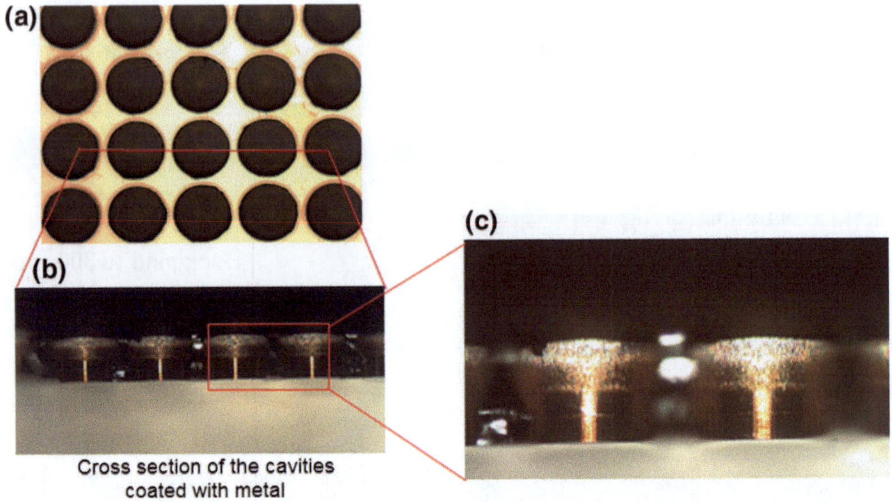

Fig. 6.4 **a** Top view of the fabricated sample coated with Cu film. **b**, **c** The cross section of the holes coated with Cu layer inside the holes

the PDMS is cured for 1 h at 120 °C. The PDMS pillars are then peeled off from the mold. The fabrication steps are shown in Fig. 6.5.

The SEM images of the fabricated PDMS pillars are shown in Fig. 6.6. As shown in the top view of PDMS pillars in Fig. 6.6a, the diameters of the base of PDMS pillar and top of pillar are different. The top circle is smaller in diameter as compared to the base circle. The shape of the PDMS pillar replicates the tilted the sidewalls of the Si holes and thus results into circular conical frustum shape as shown in the enlarged view in Fig. 6.6c. The frustum shape of pillar is utilized during the assembly of top and bottom parts, as explained in the next section.

6.2.3.3 Assembly

For the assembly of device, 100 μm spacers are cut out to create gap between the top and bottom part as shown in Fig. 6.7. As the pillars are frustum shaped, the spacer between the top and bottom part also create a gap in both flat contact area and pillar contact area (Fig. 6.2a). After placing the spacers on bottom part, PDMS pillars are aligned with the etched holes under an optical microscope using a manually controlled multi-axis stage. The optical image during the alignment of pillars with holes is shown in Fig. 6.8. Figure 6.8a shows the image of a misaligned pillar (yellow circle) relative to the hole etched in Si (red circle). The holes are manually aligned together as shown in Fig. 6.8b. After alignment, the two parts are pressed together and fixed together externally.

6.3 Experiments and Discussion

6.3.1 Tactile Sensing

As discussed in the working mechanism for the TENG, the designed structure is responsive to the normal and shear force in different ways. The device was tested for normal and sliding forces applied using human finger. For the testing, the device was connected to an oscilloscope and the voltage signal was acquired as the human finger interacted with the device. The results were processed using a low pass filter of 20 Hz to remove the unwanted signal. The results due to press and release (normal force) of the device using finger is shown in Fig. 6.9a, b. As the device is pressed, it results into contact between the PDMS layer and Cu film. The contact takes place in the flat contact area and also inside etched holes to the sidewalls. This leads to generation of surface charges due to triboelectric effect. The PDMS surface

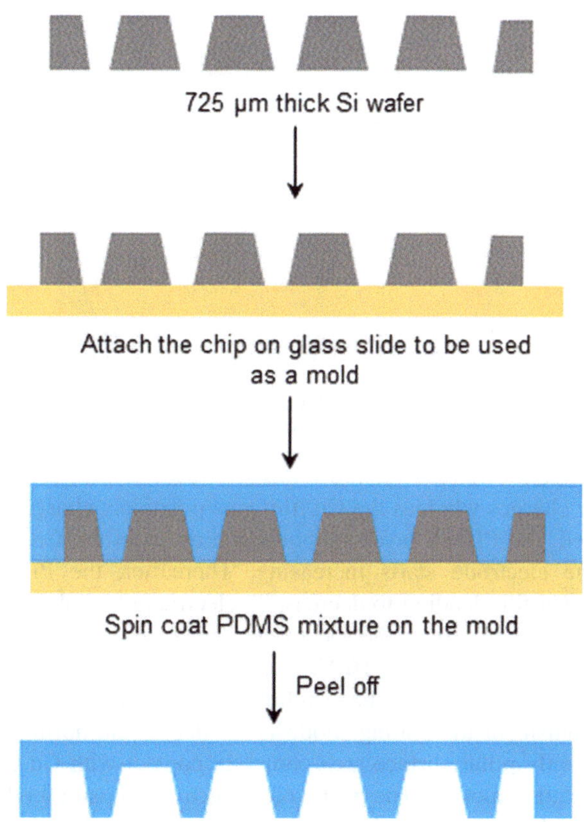

Fig. 6.5 Process flow for the fabrication of top part

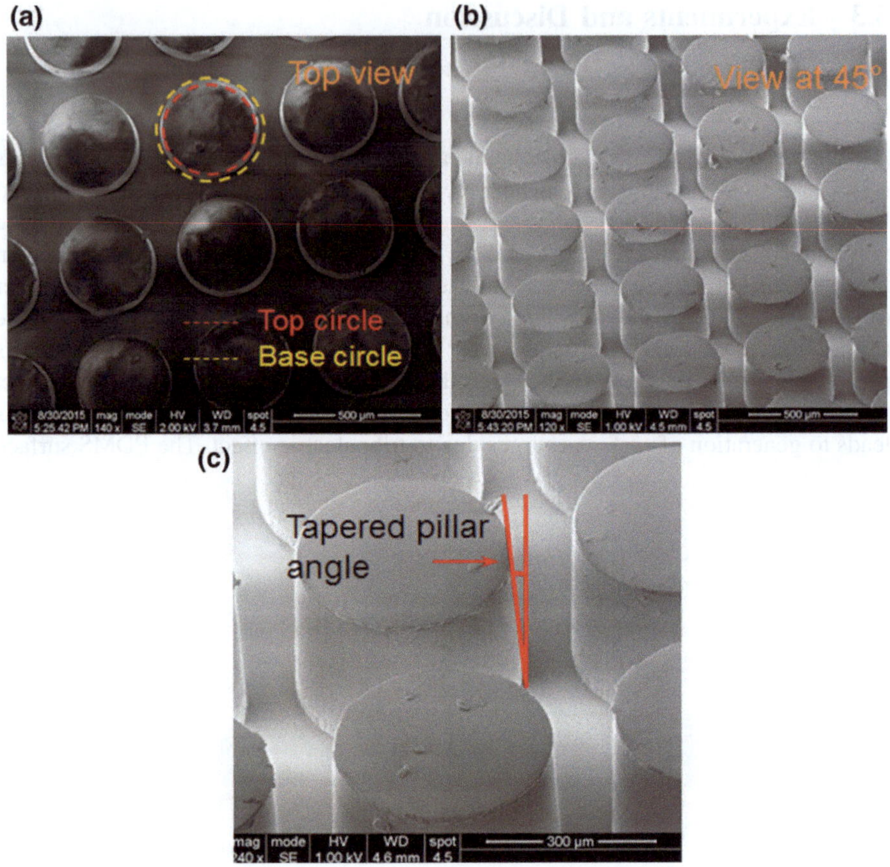

Fig. 6.6 SEM images of the fabricated PDMS pillars. **a** Top view. **b** 45° view. **c** Enlarged view of the circular conical frustum shaped pillar to show the tapered angle

gets negatively charges whereas the Cu film gets positively charged. As the normal force is released, the PDMS layer starts separating from the Cu film and electric potential at the electrode starts increasing. Thereafter, the PDMS layer again approaches the Cu film leading to decrease in electric potential. This cycle leads to electric signal as measured at the electrode. The peak voltage measured due to normal force was measured to be 16 V. The voltage signal measured due to sliding in the in-plane a direction and then sliding back, is shown in Fig. 6.9c, d. The working mechanism in the sliding mode is similar as the normal mode with the difference that only pillar surface area comes in contact with Cu coated sidewalls due to lateral relative motion. The peak voltage generated using in the sliding mode was measured to be nearly 2 V. The voltage signal generated in the normal mode

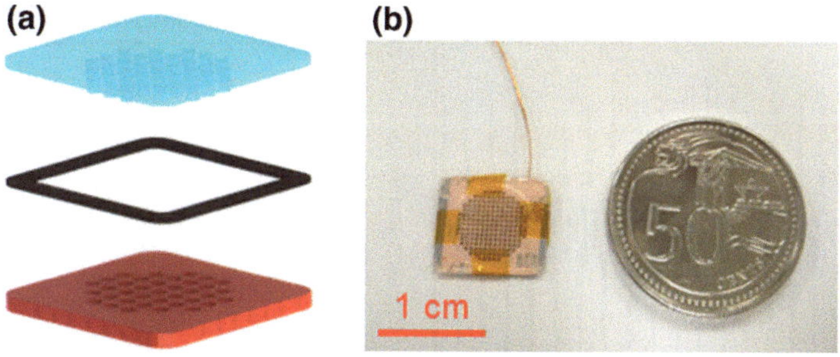

Fig. 6.7 **a** Exploded view of the device assembly. **b** Photograph of the assembled device

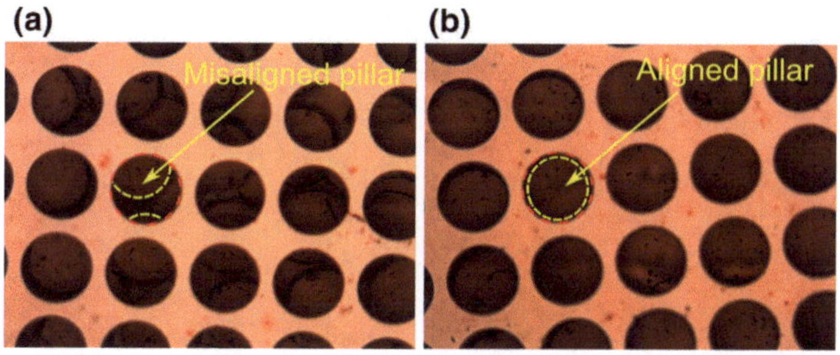

Fig. 6.8 Images of the device during alignment under optical microscope. **a** Misalignment of the PDMS pillars with Si etched holes. **b** Aligned PDMS pillars with the Si etched holes

was observed to be significantly higher than in the sliding mode. This can be explained due to significantly less contact area in the sliding mode as compared to the normal mode. The peaks in the shear mode were observed to be broader as compared to the normal mode. This was because all the PDMS pillars do not come into contact with the respective sidewalls at the same time. Due to the difference in timing of the pillars contact and separation, it leads to broader peaks. The inherent difference in the electrical response to the normal and sliding force to can be utilized to differentiate between interactions in tactile sensing.

6.3.2 Energy Harvesting Applications

The device was designed for tactile sensing with an ability to differentiate between the in-plane and out-of-plane physical interactions. But as the device is based on triboelectric mechanism and generates an electrical signal in response to mechanical

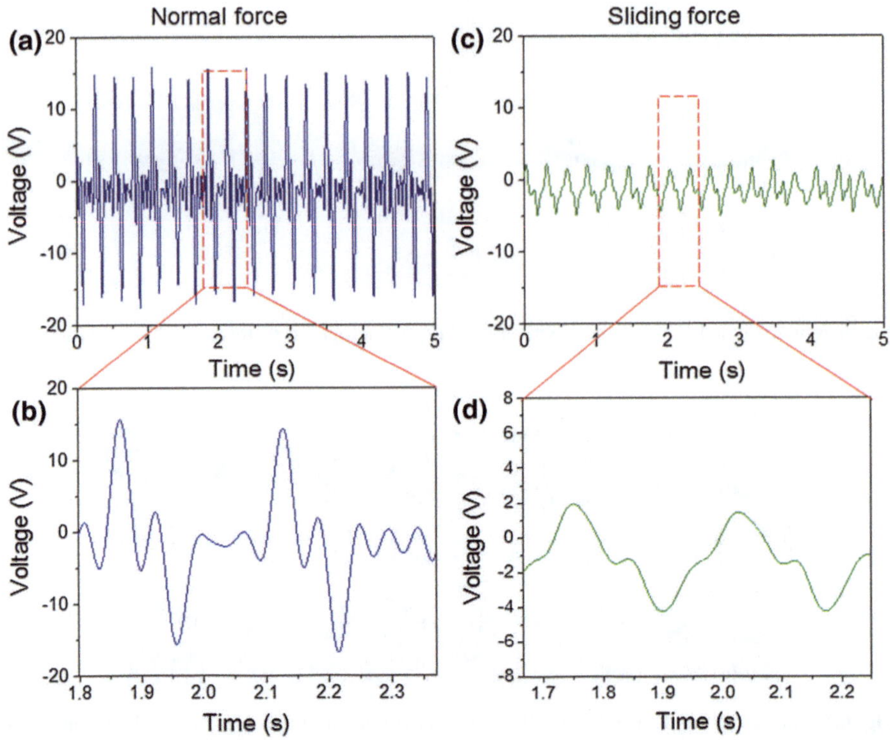

Fig. 6.9 **a**, **b** Voltage signal generated when normal force is applied on the device. **c**, **d** Voltage signal generated by due to sliding or shear force on the device

force applied, it can also be used to harvest mechanical energy from force or pressure in different directions. One of the applications of the small sized device can be in harvesting energy from typing motion on the keyboards or pressing of switches. The peak voltage and current generated due to the normal force application was measured to be ~90 V and ~0.57 μA respectively, by typing motion of the finger. Through the sliding motion, the peak voltage and current generated were measured to be ~5 V and ~0.08 μA, respectively. The generated energy was stored in a 1 μF capacitor using a full wave rectifier circuit as shown in Fig. 6.10e. The capacitor was charged up to 4 V using the typing motion of finger for about 40 s. The energy was then used to light up a red commercial LED as shown in Fig. 6.10f. The power output of the devices is limited due to the small size (1 cm × 1.2 cm of the complete device) but proposed device design can also be used for scaled up devices to improve the power output.

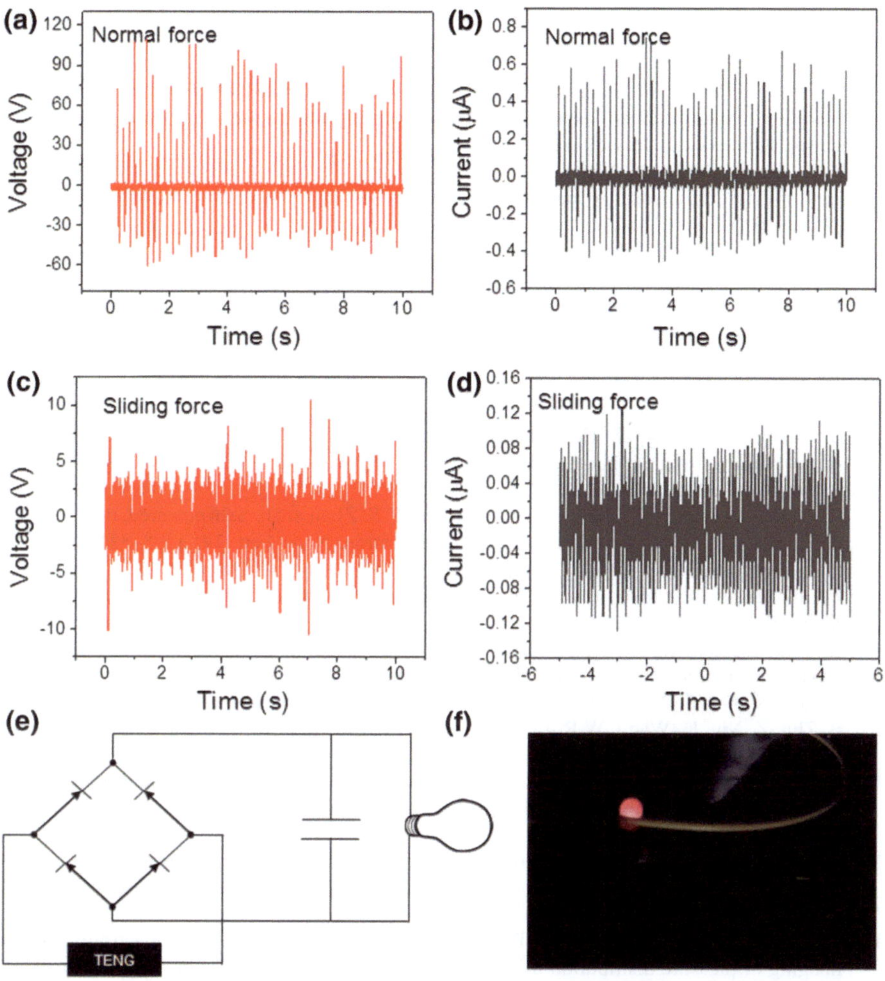

Fig. 6.10 **a** Voltage and **b** current generated from the normal force applied on the device. **c** Voltage and **d** current generated due to sliding force applied on the device. **e** Full wave rectifier circuit used to store energy in a 1 μF capacitor. **f** A *red* commercial LED lighted up using the energy stored in the capacitor after 40 s of normal force application on the device using a human finger

6.4 Summary

In this chapter, a novel mechanism based device was designed and fabricated to for tactile sensing. The mechanism can be used to differentiate between the mechanical interactions in the normal and sliding directions. The device was also demonstrated to harvest mechanical energy from mechanical force in multiple directions.

References

1. J. Chen, G. Zhu, W. Yang, Q. Jing, P. Bai, Y. Yang et al., Harmonic-resonator-based triboelectric nanogenerator as a sustainable power source and a self-powered active vibration sensor. Adv. Mater. **25**, 6094–6099 (2013)
2. Z.-H. Lin, G. Cheng, W. Wu, K.C. Pradel, Z.L. Wang, Dual-mode triboelectric nanogenerator for harvesting water energy and as a self-powered ethanol nanosensor. ACS Nano **8**, 6440–6448 (2014)
3. H. Zhang, Y. Yang, T.-C. Hou, Y. Su, C. Hu, Z.L. Wang, Triboelectric nanogenerator built inside clothes for self-powered glucose biosensors. Nano Energy **2**, 1019–1024 (2013)
4. Z. Wen, J. Chen, M.-H. Yeh, H. Guo, Z. Li, X. Fan et al., Blow-driven triboelectric nanogenerator as an active alcohol breath analyzer. Nano Energy **16**, 38–46 (2015)
5. L. Lin, Y. Xie, S. Wang, W. Wu, S. Niu, X. Wen et al., Triboelectric active sensor array for self-powered static and dynamic pressure detection and tactile imaging. ACS Nano **7**, 8266–8274 (2013)
6. B. Meng, W. Tang, Z.-H. Too, X. Zhang, M. Han, W. Liu et al., A transparent single-friction-surface triboelectric generator and self-powered touch sensor. Energy Environ. Sci. **6**, 3235–3240 (2013)
7. G. Zhu, W.Q. Yang, T. Zhang, Q. Jing, J. Chen, Y.S. Zhou et al., Self-powered, ultrasensitive, flexible tactile sensors based on contact electrification. Nano Lett. **14**, 3208–3213 (2014)
8. S.C. Mannsfeld, B.C. Tee, R.M. Stoltenberg, C.V.H. Chen, S. Barman, B.V. Muir et al., Highly sensitive flexible pressure sensors with microstructured rubber dielectric layers. Nat. Mater. **9**, 859–864 (2010)
9. H. Muhammad, C. Oddo, L. Beccai, C. Recchiuto, C. Anthony, M. Adams et al., Development of a bioinspired MEMS based capacitive tactile sensor for a robotic finger. Sens. Actuators, A **165**, 221–229 (2011)
10. B. Zhu, Z. Niu, H. Wang, W.R. Leow, H. Wang, Y. Li et al., Microstructured graphene arrays for highly sensitive flexible tactile sensors. Small **10**, 3625–3631 (2014)
11. M. Shimojo, A. Namiki, M. Ishikawa, R. Makino, K. Mabuchi, A tactile sensor sheet using pressure conductive rubber with electrical-wires stitched method. Sens. J., IEEE **4**, 589–596 (2004)
12. C. Li, P.-M. Wu, S. Lee, A. Gorton, M.J. Schulz, C.H. Ahn, Flexible dome and bump shape piezoelectric tactile sensors using PVDF-TrFE copolymer. J. Microelectromech. Syst. **17**, 334–341 (2008)
13. C. Pan, L. Dong, G. Zhu, S. Niu, R. Yu, Q. Yang et al., High-resolution electroluminescent imaging of pressure distribution using a piezoelectric nanowire LED array. Nat. Photonics **7**, 752–758 (2013)

Chapter 7
Conclusion and Recommendations for Future Work

7.1 Conclusions

The objective of thesis was to explore the applications of triboelectric effect based contact electrification for mechanical energy harvesting and self-powered sensing. A cantilever based design was fabricated to harvest low-frequency vibrations with wideband operating characteristics. The design was also used to study the effect of surface topography on the performance of TENG. Patterned layers were found to have better performance than unpatterned triboelectric layers. It was found that introduction of air voids with an optimum size was necessary to achieve optimum performance for TENG. It was also concluded that the performance of the TENGs could be significantly improved using smaller sized structures on the triboelectric layers.

To develop wearable energy harvesting design, human skin was proposed and demonstrated as a potential triboelectric layer for biomechanical energy harvesting. The design for triboelectric energy harvester was also developed into a capacitance based sensor to accurately capture the finger movement. This triboelectric nanogenerator integrated with sensor can be used for osteoarthritis rehabilitation and advanced gesture input devices.

A novel device structure was proposed for triboelectric mechanism based bidirectional tactile sensing and energy harvesting. These devices can potentially be used in keyboards and switches, which can harvest energy from the keystrokes.

For large scale fabrication of triboelectric sensors and nanogenerators, a scalable fabrication process was developed. The device used large sized polymer films patterned using roll-to-roll UV embossing to improve the device performance. The fabrication process flow was also used to develop a large size pressure sensor array for applications in posture monitoring, security systems and patient monitoring.

L. Dhakar, *Triboelectric Devices for Power Generation and Self-Powered Sensing Applications*, Springer Theses,
DOI 10.1007/978-981-10-3815-0_7

7.2 Future Work

Any potentially viable technology/product goes through various stages of development and iterations before it can be presented in a mature form so as to impact the lives of people. The future research directions and tasks to improve the triboelectric energy harvesting technology are proposed here.

7.2.1 High Current Output Triboelectric Nanogenerator

Triboelectric mechanism for energy harvesting and self-powered sensors has come a long way after it was first proposed by Prof. Z.L. Wang in 2012. Although the voltage and current output have improved by orders of magnitude since then, low current output of triboelectric mechanism based devices still remains one of the big challenges. The main reason for low current of triboelectric devices is high impedance of triboelectric mechanism based devices. The materials and device structures need to be optimized in future to reduce the device impedance and improve the current output of nanogenerators.

7.2.2 Power Management Electronics

The power management circuit to condition the output of TENG is another strategy to utilize the converted mechanical energy into electrical energy more efficiently. The high voltage and low current, output can be converted to voltage suitably low for commercial electronic devices and high current sufficient to power them.

7.2.3 Contributions

This thesis has made significant contributions for the design, application and potentially feasible process for large volume manufacturing of triboelectric energy harvesting and self-powered sensors. These are listed below:

- A study has been conducted about the effect of surface topography on the performance of the triboelectric mechanism based devices. This study is of significance to set the design rules for triboelectric nanogenerators and self-powered sensors [1, 2].
- Human skin has been proposed and demonstrated as a triboelectric material for wearable TENGs. This research work is an important step towards powering body wearable devices in future [3].

- A novel sensing mechanism based on capacitance change between human skin and device electrode, has been demonstrated for detection of human activity [4].
- A first time demonstration has been done for scalable fabrication of TENGs. These fabrication steps have also been demonstrated for the fabrication of large scale sensors arrays based on the triboelectric mechanism. This work has contributed towards low cost and high throughput production of energy harvesters and self-powered sensors [5].

The complete list of publications led by the research conducted during the Ph.D. work is listed in references.

References

1. L. Dhakar, F. Tay, C. Lee, Investigation of contact electrification based broadband energy harvesting mechanism using elastic PDMS microstructures. J. Micromech. Microeng. **24**, 104002 (2014)
2. L. Dhakar, F.E.H. Tay, C. Lee, Development of a broadband triboelectric energy harvester with SU-8 micropillars. J. Microelectromech. Syst. **24**, 91–99 (2015)
3. L. Dhakar, F. Tay, C. Lee, Skin Based Flexible Triboelectric Nanogenerators with Motion Sensing Capability, in *28th IEEE International Conference on Micro Electro Mechanical Systems (MEMS), 2015*, 2015, pp. 106–109
4. L. Dhakar, P. Pitchappa, F.E.H. Tay, C. Lee, An intelligent skin based self-powered finger motion sensor integrated with triboelectric nanogenerator. Nano Energy **19**, 532–540 (2015)
5. L. Dhakar, S. Gudla, X. Shan, Z. Wang, F. Tay, C.-H. Heng, et al., Large scale triboelectric nanogenerator and self-powered pressure sensor array using low cost roll-to-roll UV embossing. Scientific Reports **6**, 22253 (2016)

Appendix
Modeling Equations for Finger Movement Using Joint Angle and Length Between Joint

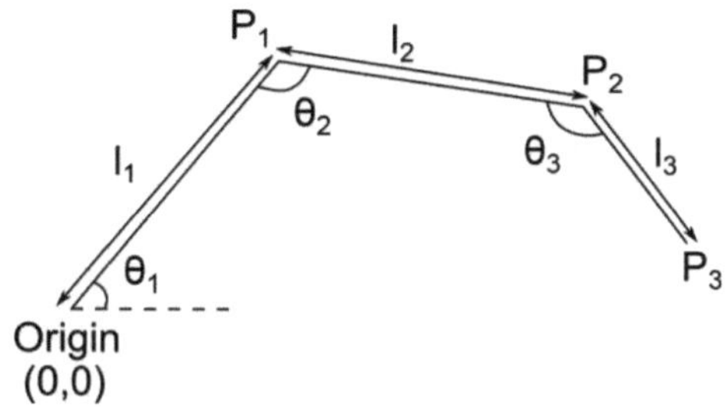

Fig. A.1 Demonstration of algorithm for human machine-interfacing or advanced input methods. Schematic of finger to calculate the coordinates of finger. Sensors are attached at Origin, P1 and P2 which can measure the angles. l_1, l_2 and l_3 are known parameters

Co-ordinates of the points (Fig. A.1):

P1: $x = l_1 \cos(\theta_1)$, $y = l_1 \sin(\theta_1)$
P2: $x = l_1 \cos(\theta_1) - l_2 \cos(\theta_1 + \theta_2)$, $y = l_1 \sin(\theta_1) - l_2 \sin(\theta_1 + \theta_2)$
P3: $x = l_1 \cos(\theta_1) - l_2 \cos(\theta_1 + \theta_2) + l_3 \cos(\theta_1 + \theta_2 + \theta_3)$, $y = l_1 \sin(\theta_1)$

$$-l_2 \sin(\theta_1 + \theta_2) + l_3 \sin(\theta_1 + \theta_2 + \theta_3)$$

Variation ranges:

$$\theta_1 \in [90°, 180°]$$
$$\theta_2 \in [60°, 180°]$$
$$\theta_3 \in [90°, 180°]$$

In the equations points P1, P2 are the points at joints and point P3 is the finger tip. It is assumed the length between the finger joints l_1, l_2 and l_3 are constant. The movement of the finger has been simulated in **Video 1** for finger movement using above equations and the data acquired from sensor for finger joint.

L. Dhakar, *Triboelectric Devices for Power Generation and Self-Powered Sensing Applications*, Springer Theses,
DOI 10.1007/978-981-10-3815-0

MIX
Papier aus verantwortungsvollen Quellen
Paper from responsible sources
FSC® C105338

If you have any concerns about our products,
you can contact us on
ProductSafety@springernature.com
In case Publisher is established outside the EU,
the EU authorized representative is:
Springer Nature Customer Service Center GmbH
Europaplatz 3, 69115 Heidelberg, Germany
Printed by Libri Plureos GmbH
in Hamburg, Germany